Lobachevski Illuminated

Nikolai Ivanovich Lobachevski (1792–1856)

AMS/MAA | SPECTRUM

VOL **69**

Lobachevski Illuminated

Seth Braver

Providence, Rhode Island

Originally published by
The Mathematical Association of America, 2011.
ISBN: 978-1-4704-5640-5
LCCN: 2010939697

SPECTRUM SERIES

The Spectrum Series of the Mathematical Association of America was so named to reflect its purpose: to publish a broad range of books including biographies, accessible expositions of old or new mathematical ideas, reprints and revisions of excellent out-of-print books, popular works, and other monographs of high interest that will appeal to a broad range of readers, including students and teachers of mathematics, mathematical amateurs, and researchers.

Contents

Introduction

Through the ostensibly infallible process of logical deduction, Euclid of Alexandria (ca. 300 B.C.) derived a colossal body of geometric facts from a bare minimum of genetic material: five *postulates* — five simple geometric assumptions that he listed at the beginning of his masterpiece, the *Elements*. That Euclid could produce hundreds of unintuitive theorems from five patently obvious assumptions about space, and, still more impressively, that he could do so in a manner that precluded doubt, sufficed to establish the *Elements* as mankind's greatest monument to the power of rational organized thought. As a logically impeccable, tightly wrought description of space itself, the *Elements* offered humanity a unique anchor of definite knowledge, guaranteed to remain eternally secure amidst the perpetual flux of existence — a rock of certainty, whose truth, by its very nature, was unquestionable.

This universal, even transcendent, aspect of the *Elements* has profoundly impressed Euclid's readers for over two millennia. In contrast to all explicitly advertised sources of transcendent knowledge, Euclid never cites a single authority and he never asks his readers to trust his own ineffably mystical wisdom. Instead, we, his readers, need not accept anything on faith; we are free and even encouraged to remain skeptical throughout. Should one doubt the validity of the Pythagorean Theorem (*Elements* I.47), for example, one need not defer to the reputation of "the great Pythagoras". Instead, one may satisfy oneself in the manner of Thomas Hobbes, whose first experience with Euclid was described by John Aubrey, in his *Brief Lives*, in the following words.

> He was 40 years old before he looked on Geometry; which happened accidentally. Being in a Gentleman's library, Euclid's *Elements* lay open, and 'twas the 47 *E. Libri I*. He read the Proposition. *By G* —, says he, (he would now and then sweare an emphaticall Oath by way of emphasis) *this is impossible*! So he reads the Demonstration of it, which referred him back to such a Proposition; which proposition he read. That referred him back to another, which he also read. *Et sic deincips* and so on that at last he was demonstratively convinced of that truth. This made him in love with Geometry.

The *Elements* was an educational staple until the early twentieth century. So long as reading it remained a common experience among the educated, Euclid's name was synonymous with demonstrable truth[1]. It is not an exaggeration to assert that Euclid was the envy of both philosophy

---

[1] At the very least, the demonstrations in the *Elements* were acknowledged as the strongest possible sort of which the rational mind is capable. "It is curious to observe the triumph of slight incidents over the mind: — What incredible weight they have in forming and governing our opinions, both of men and things — that trifles, light as air, shall waft a belief into the soul, and plant it so immovably within it – that Euclid's demonstrations, could they be brought to batter it in breach, should not all have power to overthrow it." (Laurence Sterne, *The Life and Opinions of Tristram Shandy.* Book IV, Ch. XXVII)

and theology. In his *Meditations*, Descartes went so far as to base his certainty that God exists on his certainty that Euclid's 32nd proposition is true. This was but a single instance out of many in which theology has tried to prop itself up against the rock of mathematics. Euclid's *Elements*, for all its austerity, appeals to a deep-seated human desire for certainty. This being the case, any individual with the impertinence to challenge Euclid's authority was certain to inspire reactions of both incredulity and scorn.

But how exactly *can* one challenge Euclid's authority? Euclid asks us to accept nothing more than five postulates, and all else follows from pure logic. Therefore, if there is anything to challenge in the *Elements*, it can only be in the postulates themselves. The first four seem almost too simple to question. Informally, they describe the geometer's tools: a straightedge, a compass, and a consistent means for measuring angles. The fifth postulate, however, is of a rather different character:

> That, if a straight line falling on two straight lines make the interior angles on the same side less than two right angles, the two straight lines, if produced indefinitely, meet on that side on which are the angles less than the two right angles.

This is Euclid's famous *parallel postulate*, so called because it forms the basis for his theory of parallels, which, in turn, forms the basis for nearly everything else in geometry. Modern geometry texts almost invariably replace this postulate with an alternative, to which it is logically equivalent: *given a line and a point not on it, there is exactly one line that passes through the point and does not intersect the line*. Particularly when expressed in this alternate form, the parallel postulate does strike most as "self-evident", and thus beyond question for any sane individual. It would seem, therefore, that Euclid has no significant weaknesses; his geometry is *the* geometry — impregnable, inevitable, and eternal.

The timeless, almost icy, perfection that characterizes Euclid's work made it not only a logical masterpiece, but an artistic one as well. In this latter aspect, commentators often singled out the parallel postulate as the unique *aesthetic* flaw in the *Elements*. The problem was that the parallel postulate seemed out of place: it read suspiciously like a theorem — something that Euclid should have *proved* from his earlier postulates, instead of adjoining it to their ranks. This structural incongruity — a postulate that "should be" a theorem — disturbed many mathematicians from antiquity to the 19th century. We may safely presume that Euclid tried and failed to prove the postulate as a theorem. We *know* that Euclid's followers and admirers tried to do as much, hoping to perfect their master's work by polishing away this one small but irritating blemish. Many believed that they had succeeded.

Records of flawed "proofs" rarely survive, as there generally seems no reason to preserve them, so the astonishing number of alleged proofs of the parallel postulate that have come down to us should serve to indicate just how much attention was given to this problem. Proclus, a 5th-century neo-Platonic philosopher, who wrote an extensive commentary on the first book of the *Elements*, describes two attempts: one by Posidonius (2nd century B.C.), the other by Ptolemy (the 2nd-century A.D. author of the *Almagest*, the Bible of geocentric astronomy). Both arguments, Proclus points out, are inadmissible because they contain subtle flaws. After detailing these flaws, Proclus proceeded to give his own proof, thus settling the matter for once and all — or so he thought. Proclus' proof, for all his critical acumen, was just as faulty as those he had criticized.

We have flawed proofs by Aghanis (5th century) and Simplicius (6th-century), two Byzantine scholars. Many others by medieval Islamic mathematicians have survived, including attempts by al-Jawhari and Thabit ibn Qurra in the 9th century, al-Haytham and Omar Khayyám in the 11th, and Nasir-Eddin al-Tusi in the 13th. There are even a few specimens from medieval Europe, such as those conceived by Vitello in the 13th century and Levi ben Gerson in the 14th. A veritable horde of later Europeans left purported proofs of the postulate (to cite just a few examples: Christopher Clavius in 1574, Pietro Antonio Cataldi in 1604, Giovanni Alfonso Borelli in 1658, Gerolamo Saccheri in 1733, Louis Bertrand in 1788, and Adrien Marie Legendre, who published many attempts between 1794 and 1832). Indeed, in 1763, G.S. Klügel wrote a dissertation examining no less than twenty-eight unsound "proofs" of the postulate. Interestingly, most would-be postulate provers followed Proclus in explicitly criticizing one or more of their predecessors' attempts before giving their own flawed "proof to end all proofs".[2]

Adhering to long-standing custom, Nikolai Ivanovich Lobachevski (1792–1856) began many of his own works on the subject by criticizing the alleged proofs of his immediate predecessor, Legendre. However, instead of forging the chain's next link, Lobachevski suggested that the chain be discarded altogether. He insisted that the parallel postulate *cannot* be proved from Euclid's first four postulates. In this sense, Lobachevski was a great defender of Euclid: he felt that Euclid was fully justified in assuming the parallel postulate as such; indeed, he believed that Euclid had no other way to obtain it.

In another sense, Lobachevski believed that Euclid was wholly *unjustified* in assuming the parallel postulate, for we cannot be certain that it accurately describes the behavior of lines in physical space. Euclidean tradition declares that it does, but the universe is not obliged to respect humanity's traditional beliefs about space, even those codified by its great authority, Euclid of Alexandria. Lobachevski considered the validity of the parallel postulate an empirical question, to be settled, if possible, by astronomical measurements.

Unorthodoxy quickly led to heresy: proceeding from the assumption that the parallel postulate does *not* hold, Lobachevski began to develop a *new* geometry, which he called *imaginary geometry*[3], whose results contradicted Euclid's own. He first described this strange new world on February 24, 1826, in a lecture at the University of Kazan. His first written publication on the subject dates from 1829. Several others followed, and after a decade of failed attempts to convince his fellow Russians of the significance of his work, he published accounts of it in French (in 1837) and German (in 1840), hoping to attract attention in Western Europe. He found none. By the time that he wrote *Pangeometry* (1855), he was blind (he had to dictate the book), exhausted, and embittered. He died the following year.[4]

In fact, although Lobachevski never knew it, his work did find one sympathetic reader in his lifetime: Karl Friedrich Gauss (1777–1855), often classed with Isaac Newton and Archimedes as one of the three greatest mathematicians who have ever lived. Gauss shared Lobachevski's convictions regarding the possibility of an alternate geometry, in which the parallel postulate does

---

[2] For detailed descriptions of many alleged proofs of the postulate, consult Rosenfeld (Chapter 2) or Bonola (Chapters 1 and 2).

[3] By the end of his life, he preferred the name *pangeometry*, for reasons that will become clear by the end of *The Theory of Parallels*. Other common adjectives for Lobachevski's geometry are *non-Euclidean* (used by Gauss), *hyperbolic* (introduced by Felix Klein), and *Lobachevskian* (used by Russians).

[4] His French paper of 1837, *Géométrie Imaginaire*, appeared in August Crelle's famous journal, *Journal für die Reine und Angewandte Mathematik* (Vol. 17, pp. 295–320). His German publication of 1840 was *The Theory of Parallels*; its full title is *Geometrische Untersuchungen zur Theorie der Parallelinien* (*Geometric Investigations on the Theory of Parallels*). Lobachevski wrote two versions of *Pangeometry*, one in French and one in Russian.

not hold. He reached these conclusions earlier than Lobachevski, but abstained, very deliberately, from publishing his opinions or investigations. Fearing that his ideas would embroil him in controversy, the very thought of which Gauss abhorred, he confided them only to a select few of his correspondents, most of them astronomers. When Gauss read an unfavorable review of Lobachevski's *Theory of Parallels*, he dismissed the opinions of the reviewer, hastened to acquire a copy of the work, and had the rare pleasure of reading the words of a kindred, but more courageous, spirit. Gauss was impressed; he even sought out and read Lobachevski's early publications in Russian. To H.C. Schumacher, he wrote in 1846, "I have not found anything in Lobachevski's work that is new to me, but the development is made in a different way from the way I had started and, to be sure, masterfully done by Lobachevski in the pure spirit of geometry."

True to his intent, Gauss' radical thoughts remained well-hidden during his lifetime, but within a decade of his death, the publication of his correspondence drew the attention of the mathematical world to non-Euclidean geometry. Though the notion that there could be two geometries did indeed generate controversy, the fact that Gauss himself endorsed it was enough to convince several mathematicians to track down the works of the unknown Russian whom Gauss had praised so highly. Unfortunately, Lobachevski reaped no benefit from this interest; he was already dead by that time, as was the equally obscure Hungarian mathematician, János Bolyai (1802–1860), whose related work also met with high praise in Gauss' correspondence.

Bolyai had discovered and developed non-Euclidean geometry independently of both Lobachevski and Gauss. He published an account of the subject in 1832, but it had essentially no hope of finding an audience: it appeared as an appendix to a two-volume geometry text, written by his father, Farkas Bolyai, in Latin. Farkas Bolyai, who had known Gauss in college, sent his old friend a copy of his son's revolutionary studies. Gauss' reply — that all this was already known to him — so discouraged the young János, that he never published again, and even ceased communicating with his father, convinced that he had allowed Gauss to steal and take credit for his own discoveries. Father and son were eventually reconciled, but Bolyai was doubly disheartened some years later to learn that his own *Appendix* could not even claim the honor of being the first published account of non-Euclidean geometry: Lobachevski's earliest Russian paper antedated it by several years.

As mathematicians began to re-examine the work of Lobachevski and Bolyai, translating it into various languages, extending it, and grappling with the philosophical problems that it raised, they changed the very form of the subject in order to assimilate it into mainstream mathematics. By 1900, non-Euclidean geometry remained a source of wonder, but it had ceased to be a controversial subject among mathematicians, who were now describing it in terms of differential geometry, projective geometry, or Euclidean "models" of the non-Euclidean plane. These developments and interpretations helped mathematicians domesticate the somewhat nightmarish creatures that Lobachevski and Bolyai had loosed upon geometry. Much was gained, but something of great psychological importance was also lost in the process. The tidy forms into which the subject had been pressed scarcely resembled the majestic full-blooded animal that Lobachevski and Bolyai had each beheld, alone, in the deep dark wild wood.

Today, in 2011, the vigorous beast is almost never seen in its original habitat. Just as we give toy dinosaurs and soft plushy lions to children, we give harmless non-Euclidean toys, such as the popular Poincaré disc model, to mathematics majors. We take advanced students to the zoo of differential geometry and while we are there, we pause — briefly, of course — to point out a captive specimen of hyperbolic geometry, sullenly pacing behind bars of constant negative curvature.

If we are to understand the meaning of non-Euclidean geometry — to understand why it wrought such important changes in mathematics — we must first recapture the initial fascination and even the horror that mathematicians felt when confronted with the work of Lobachevski and Bolyai. This, however, is difficult. The advent of non-Euclidean geometry changed the mathematical landscape so profoundly that the pioneering works themselves were obscured in the chaos of shifting tectonic plates and falling debris. Mathematical practices of the early 19th century are not the same as those of the early 21st. The gap of nearly two centuries generally precludes the possibility of a sensitive reading of Lobachevski's works by a modern reader. This book is an attempt to rectify the situation, by supplying the contemporary reader with all of the tools necessary to unlock this rich, beautiful, but generally inaccessible world. But where does one start?

Gauss left us nothing to work with. Bolyai's *Appendix* is out of the question; his writing is often terse to the point of incomprehensibility. Lobachevski is far clearer, but he too makes heavy demands on his readers. Perhaps we should read his earliest works? In 1844, Gauss described them (in a letter to C.L. Gerling) as "a confused forest through which it is difficult to find a passage and perspective, without having first gotten acquainted with all the trees individually." At the other chronological extreme, Lobachevski's final work, *Pangeometry*, is inappropriate for beginners since it merely summarizes the elementary parts of the subject, referring the reader to *The Theory of Parallels*, his German book of 1840, for proofs. *Pangeometry* does make a logical second book to read, but the book that it leans upon, *The Theory of Parallels*, remains the best point of ingress for the modern mathematician.

Accordingly, the following pages contain a new English version of *The Theory of Parallels*, together with mathematical, historical, and philosophical commentary, which will expand and explain Lobachevski's often cryptic statements (which even his contemporaries failed to grasp), and link his individual propositions to the related work of his predecessors, contemporaries, and followers. Resituated in its proper historical context, Lobachevski's work should once again reveal itself as an exciting, profound, and revolutionary mathematical document.

The complete text of Lobachevski's *Theory of Parallels* appears **twice** within the pages of this book. In the appendix, it appears as a connected whole, in its first English translation since Halsted's in 1891. In the body of the book, the complete text appears a second time, but broken into more than 100 pieces; I have woven my illumination around these hundred-odd pieces. Lest there be any confusion as to whose voice is speaking at any given place in the book, Lobachevski's words have been printed in red, while everything else is printed in black.

A Note to the Reader

"Begin at the beginning," the King said, very gravely,
"and go on till you come to the end: then stop."
—*Alice in Wonderland*, Chapter 12

Although following the King of Hearts' advice may be the most rigorous way to read *Lobachevski Illuminated*, it is hardly the only way. Beginning at the beginning is always a sensible idea, but one need not feel compelled to "go on" through the details of each and every auxiliary proof that I provide along the way. Many readers, first-time readers in particular, will simply want an overview of Lobachevski's accomplishments and methods. If you are such a person, then you should feel free to skip any technical proofs that threaten to divert you from the main narrative thread.

Ideally, one should at least read the *statements* of the propositions that I prove in the notes. What one chooses to do with them will then vary from reader to reader. Some will want to try coming up with their own proofs. Others will simply read mine. Still others will take the statements on faith and move on, confident in the knowledge that the proofs are there, patiently waiting, should they ever need to be consulted. All of these are reasonable approaches.

Of course, the further one travels into the counterintuitive non-Euclidean countryside, the less confidence one will have in dismissing anything as "obvious", "trivial", or "a mere technicality". If you have never left the Euclidean world before, then be forewarned: you are about to embark on a thoroughly disorienting (but strangely exhilarating) journey. Some readers will be more comfortable taking one tiny step at a time into this new land, mapping the terrain slowly and carefully, proving everything in detail, until even its most alien features take on a kind of unexpected familiarity. Others will charge boldly ahead, skipping many proofs, eager to reach the dark heart of the matter as quickly as possible; they will get there, of course, but are likely to find themselves so thoroughly befuddled that they will almost certainly want to go back and carefully retrace their steps so as to make some retrospective sense of the strange sights they have beheld. Again, both approaches are fine, and are ultimately a matter of individual psychology. An approach somewhere between these two extremes is probably best.

Of course, there will be some readers who will want considerably *more* technical detail than I've provided. In particular, some may desire a rigorously-argued Hilbert-style examination of the foundations of geometry. Others won't stop even there, and will want to dig into the primal matter of logic itself. But as with its readers, so a book's author must draw the line somewhere.

I have laid out the meal. Fall to it and eat. Only you can decide what to put on your plate.

Acknowledgements

Like everything else in the universe, *Lobachevski Illuminated* is a product of chance. Had any of my stone-age ancestors died before begetting offspring, I never would have existed. You might not have either. At any rate, you would not be reading *Lobachevski Illuminated*. If describing the full web of coincidences that ultimately led to this book would be tantamount to narrating the history of the universe, the least I can do is to tip my hat to a few people who loom particularly large in the book's genesis.

Hats off to Richard Mitchell at UC-Santa Cruz. His profound yet playful knowledge of geometry and history, his collection of outrageously complicated polyhedra, his wry skepticism and sense of humor all suggested to me an approach to life and to mathematics far saner than that which seems to engulf the typical harried academic[5]. When, as a master's student at UCSC, I was becoming thoroughly disoriented by the nested abstractions of modern mathematics, Richard encouraged me to delve into the subject's history as a means of recovering the basic intuitions that motivated such abstractions in the first place. I acknowledge you, O Richard!

At the University of Montana, where I wrote the dissertation that became the basis of this book, I was blessed with two excellent advisors, Greg St. George and Karel Stroethoff. Over the years, I have met hundreds of professionals who publish mathematics papers and teach mathematics for a living. The number of actual *mathematicians* I've met, however, I can count on one hand. True mathematicians are a rare breed; I feel privileged to have worked with Karel, who is one of them. I worked with Greg from the moment I arrived at UM. Had I not found such a humane advisor, to whom I could speak as friend to friend rather than as apprentice to master, I probably would have quit the doctoral program. Greg had the wisdom to let me pursue whatever mathematical topic I felt interested in, however unfashionable, which we would then discuss at our informal weekly meetings – meetings that, as often as not, evolved into discussions of languages, philosophy, and literature. He was never one to respect the artificial boundaries between "disciplines", for which failing I extend my humble thanks.

And in a final flourish... thanks to Gerald Alexanderson, editor of the MAA's Spectrum series, for his generous support of this book; to Llyd Wells for his friendship and post-seminar whiskey; to Reb Hastrev for his sparkling wit and iconoclastic spirit; to Shannon Michael for this and that; to Professor Q.A. Wagstaff for his luminous discourse on modern education; to Bernard Russo for his curious tract, *Group Interaction Diagrams*; to Edward Lear, whom it is pleasant to know; to Lucretius and Ovid; to Samuel Beckett; to Edwin Arlington Robinson; to Emily

---

[5] My brain has often been gripped by visions of university professors as members of a weird chain gang condemned to perform meaningless tasks over and over again, but instead of breaking large rocks into smaller rocks, these prisoners endlessly write and publish papers that no one will ever read.

Dickinson; to the broad-shouldered son of Ariston; to Flann O'Brien; to Bohuslav Martinů; to György Ligeti; to Percival Bartlebooth for the watercolors; to my dear Uncle Toby for the lovely map of Namur; to Tom Lehrer for two portraits of Lobachevski; to V. Mora, for demonstrating that last names (in ancient Greek) are often prophetic; to N. Artemiadis, for his letter of explanation; to Vanni Fucci, for the delicious figs; to Anna Livia Plurabelle, whose leaves are drifting from her; and to Bartleby, because he is so very good.

Theory of Parallels—Lobachevski's Introduction

In geometry, I have identified several imperfections, which I hold responsible for the fact that this science, apart from its translation into analysis, has taken no step forward from the state in which it came to us from Euclid. I consider the following to be among these imperfections: vagueness in the basic notions of geometric magnitudes, obscurity in the method and manner of representing the measurements of such magnitudes, and finally, the crucial gap in the theory of parallels. Until now, all mathematicians' efforts to fill this gap have been fruitless. Legendre's labors in this area have contributed nothing. He was forced to abandon the one rigorous road, turn down a side path, and seek sanctuary in extraneous propositions, taking pains to present them—in fallacious arguments—as necessary axioms.

I published my first essay on the foundations of geometry in the "Kazan Messenger" in the year 1829. Hoping to provide an essentially complete theory, I then undertook an exposition of the subject in its entirety, publishing my work in installments in the "Scholarly Journal of the University of Kazan" in the years 1836, 1837, and 1838, under the title, "New Principles of Geometry, with a Complete Theory of Parallels". Perhaps it was the extent of this work that discouraged my countrymen from attending to its subject, which had ceased to be fashionable since Legendre. Be that as it may, I maintain that the theory of parallels should not forfeit its claim to the attentions of geometers. Therefore, I intend here to expound the essence of my investigations, noting in advance that, contrary to Legendre's opinion, all other imperfections, such as the definition of the straight line, will prove themselves quite foreign here and without any real influence on the theory of parallels.

Legendre

> "I have read M. Legendre's book. Ach! It is beautiful! You shall find in it no flaw!"
> —Herr Niemand, in *Euclid and his Modern Rivals*[1].

In 1794, when Lobachevski was an infant, Adrien Marie Legendre published his famous *Éléments de Géométrie*, a textbook that attempted to improve Euclid's presentation of geometry by

---

[1] Carroll, p. 54.

simplifying the proofs in Euclid's *Elements*, and reordering its propositions. In subsequent editions and translations, Legendre's text became a 19th-century educational staple. Its admirers were legion; they taught and learned from it in locations throughout Europe, the antebellum United States, and even in Lobachevski's remote Russian city of Kazan. Legendre died in 1833, but his textbook remained immensely popular for the duration of the 19th century. In the epigraph for this section, Herr Niemand waxes enthusiastic over the 14th edition of Legendre's *Éléments*, published in 1860[2].

Legendre never doubted that the parallel postulate was a logical consequence of Euclid's first four axioms. Indeed, he claimed to have discovered several proofs of the postulate—all flawed, of course. New editions of his textbook often featured new proofs of the postulate, not because Legendre recognized that the proof in the previous edition was invalid, but rather, because he feared that the old proof had been too complex for beginners to follow. I have reproduced one of his "proofs" in the notes to TP 19. His argument is ingenious, but contains a very subtle flaw, as you shall see. Each of his proofs contains a hidden circular argument of the same variety: he implicitly assumes the truth of a property equivalent to the parallel postulate. Legendre's little loops of logic are the "side paths" to which Lobachevski refers above. In the introduction to his *New Principles of Geometry* (1835–8), Lobachevski describes and criticizes some of Legendre's attempts on the postulate[3].

Near the end of his life, Legendre summarized his work on the parallel postulate in his memoir, *Reflections on the Different Ways of Proving the Theory of Parallels or the Theorem on the Sum of the Three Angles of the Triangle*. In it, he laments the inherent difficulty of proving the postulate, and suggests reasons why it had resisted proof for so long.

> Without doubt, one must attribute to the imperfection of common language and to the difficulty of giving a good definition of the straight line the little success which geometers have had until now when they have wanted to deduce this theorem.[4]

Lobachevski responds to this in the last sentence of his introductory remarks above. The logical reasons for the parallel postulate's necessity are not, as Legendre suggests they are, deep. Rather, they are *nonexistent*. In the pages that follow, Lobachevski will dispense with the parallel postulate, accept its negation, and defiantly proceed to develop geometry anew.

---

[2] In fact, Charles Dodgson (a.k.a. Lewis Carroll) published *Euclid and his Modern Rivals* in the year 1879. Despite the lapse of 85 years since the original publication of Legendre's book, Dodgson (in the guise of his character, Minos) clearly preferred it to all the other "modern rivals" of Euclid. He describes the book and its proofs as "beautiful", "admirable", and "a model of elegance", but worries that it may be too difficult for beginners. When one reflects that today's (2011) elementary mathematics courses are never taught from books published in 1922, one can appreciate the special nature of Legendre's text.

[3] In the same place, he criticized a popular "proof" due to the Swiss mathematician Louis Bertrand in 1778. Bertrand's *reductio ad absurdum* argument involved dubious comparisons of infinite areas, a technique that Legendre also used in one of his proofs. Areas, along with lengths and volumes, are the "geometric magnitudes" to which Lobachevski alludes; he criticizes Bertrand and others for applying techniques that hold for finite figures to infinite figures, for which they may no longer be valid. A description of Bertrand's proof is in Rosenfeld (p. 102). In modified form, this proof resurfaced in *Crelle's Journal* in 1834 (the same journal that published a paper by Lobachevski in 1837!), and as late as 1913, an article in *The Mathematical Gazette* (Vol. 7, p. 136) would claim, "Bertrand of Geneva proved the parallel-axiom finally and completely."

[4] Laubenbacher & Pengelly, p. 26

Theory of Parallels—Preliminary Theorems (1–15)

Mathematical terms cannot be defined *ex nihilo*. The words that one uses in any given definition require further definitions of their own; these secondary definitions necessitate tertiary definitions; these in turn require still others. To escape infinite regress, geometers must leave a handful of so-called primitive terms undefined. These primitive terms represent the basic building blocks from which the first defined terms may be constructed. From there, one may build upward indefinitely; all subsequent development will be grounded upon the primitive terms, and circular definitions will be avoided.

Only in the late 19th-century was such clarity achieved in the foundations of geometry. Euclid never identifies his primitive terms and several of his early definitions founder in ambiguity. His vague definition of a straight line, "a line which lies evenly with the points on itself"[1] is useless from a logical standpoint: since Euclid does not tell us what "lying evenly" means, we have no way of deciding whether a given curve is straight or not. Euclid has given us a description rather than a genuine definition of a line, and as such, he has given us something that is worthless in a strict logical development of geometry.

Mathematics encompasses more than logic, however[2]. The very fact that Euclid attempts to describe a line has philosophical significance. It suggests that, for Euclid, straight lines are "out there", capable of description. It implicitly asserts that *straight lines exist independently of the mathematicians who study them*. For one who accepts this Platonic concept of geometry, the logical gaps in *The Elements* are so superficial as to scarcely merit mention. For example, Euclid does not bother to justify the obvious fact that if a straight line enters a triangle through one of its vertices, then it must exit through the opposite side. Nevertheless, he frequently uses this fact in his proofs; he *knows* that the line must exit through the opposite side. Of course, we know it as well, but how do we know it? Where is this mysterious Platonic realm and how do our minds gain access to it? Might our intuitions about it be mistaken?

We can wage long battles over such questions, but it is much more comfortable for mathematicians to retreat to philosophical positions that are easier to defend than Platonism. Behind the bunkers of formal axiomatic development, we can generally remain safe from philosophers' attacks. There are several ways to reconstruct the formal foundations of geometry, making them nearly unassailable. In his book *Foundations of Geometry* (1900), David Hilbert based the entire subject upon five primitive terms: *point*, *line*, *contain*, *between*, and *congruent*. Hilbertian

---

[1] Euclid's "line" is our "curve".

[2] But compare Bertrand Russell: "The subject of formal logic, which has now at last shown itself to be identical with mathematics..."

formalism denies that lines (or any other undefined concepts) inhabit a reality that one can contemplate outside the context of rigorous deduction. Consequently, his definitions are not intended to describe ideal objects, but rather to endow empty words with precise mathematical meanings. Euclid was content to leave certain fundamental notions on an intuitive basis, such as the simple statement about lines and triangles mentioned in the previous paragraph, but for Hilbert, the geometric atheist, such a procedure is anathema; truth is synonymous with proof. To use theorems that one cannot prove is to abandon mathematics for theology[3].

A Rough Start: TP 1–5

> ". . . the beginning. . .was without form, and void."
> —*Genesis* 1:1-2.

Much of the defensive work that went into shoring up the foundations of geometry was inspired by the shock caused by Lobachevski's non-Euclidean geometry when it became known in the late 19th century. The existence of a second geometry raises the disturbing possibility that our basic intuitions about geometry might be fallible, after all. This foundational work, however, came after Lobachevski's death, so we should not expect to see its like in *The Theory of Parallels*. Indeed, following Euclid's tradition of doubtful preliminaries, Lobachevski begins his book with five confused "theorems", four of which should certainly be demoted to the status of descriptions (or axioms), as they do not admit proof on the basis of Euclid's axioms. Even the one genuine theorem in the group (TP 4) is superfluous: it is just a special case of TP 7.

> Lest my reader become fatigued by a multitude of theorems whose proofs present no difficulties, I shall list here in the preface only those that will actually be required later.
>
> **1)** *A straight line covers itself in all its positions.* By this, I mean that a straight line will not change its position during a rotation of a plane containing it if the line passes through two fixed points in the plane.

In TP 1, Lobachevski begins with an assertion suspiciously similar to Euclid's definition of a straight line. Unlike Euclid, he seems to recognize its weakness, and tries to clarify it with his second sentence. He does not succeed. He is apparently claiming that straight lines are the fixed-point-sets of spatial rotations. To *prove* that this is so, we would have to demonstrate that the set of fixed points under such a rotation satisfies the definition of a straight line. Since Euclid's definition is clearly useless for this task (How does one demonstrate that a set of points "lie evenly with themselves?"), and Lobachevski proposes no alternate definition, we must conclude that TP 1 cannot be proved rigorously.

Just as Euclid never subsequently refers to his vague definition of a straight line, Lobachevski never refers to TP 1 elsewhere in the *Theory of Parallels*. What are we to make of this inauspicious beginning? Why does Lobachevski begin his treatise with a vague statement, labeled as a theorem yet incapable of proof, to which he never subsequently refers? That this inscrutable

---

[3] Compare Blaise Pascal's dictum: "Reason is the slow and tortuous method by which those who do not know the truth discover it."

pronouncement heads a list of theorems specifically designated as vital for the sequel and amenable to easy proof makes it stranger still.

I believe that we must read TP 1 as I have suggested that we read Euclid's definition of a straight line: as an implicit assertion that straight lines have an intrinsic "nature", reflected in our experience of straightness (or approximate straightness) in the natural world. Thus, the implicit role of this proposition is to rule out certain "unnatural" behaviors of straight lines, such as self-intersection. A curve in the plane that loops back and intersects itself obviously does not correspond to our intuition of straightness. Considered as logical tools, Euclid's definition of straightness and Lobachevski's TP 1 are undoubtedly problematic, if not altogether meaningless. If any value can be ascribed to them, it must be historical rather than mathematical. Each, if nothing else, hints at an underlying Platonist philosophy of mathematics in their respective author. This indication of a Platonist strain in Lobachevski's work helps explain why his investigations into the foundations of geometry took such a different path than those undertaken by Hilbert.

2) *Two straight lines cannot intersect one another in two points.*

Two lines intersecting one another in two points would violate Euclid's first postulate, which states that a *unique* line may be drawn through any two points.[4]

3) *By extending both sides of a straight line sufficiently far, it will break out of any bounded region. In particular, it will separate a bounded plane region into two parts.*

These exemplify the types of intuitive statements that Euclid and Lobachevski use without axiomatic justification. Naturally, Hilbert's foundations allow one to prove them, but only after devoting a good deal of labor to defining terms and establishing a host of preliminary lemmas.

In *The Theory of Parallels*, Lobachevski uses TP 3 as follows. When a line enters a bounded figure, such as a triangle, TP 3 simply guarantees that the line, if extended far enough, will eventually come out again.

We shall discuss this again in TP 17, when Lobachevski first invokes TP 3.

4) *Two straight lines perpendicular to a third will never intersect one another, regardless of how far they are extended.*

TP 4 is just a special case of TP 7.

5) *When a straight lines passes from one side to the other of a second straight line, the lines always intersect.*

Here, Lobachevski posits the *continuity* of straight lines: they lack holes through which one might thread a second line. Euclid assumes this as obvious. We shall do the same and proceed onward.

---

[4] Euclid explicitly postulates only the *existence* of a line through any two points, but his failure to mention uniqueness seems to have been an oversight, since he makes specific use of its uniqueness several times (in his proof of I.4, for example). Many editions of the *Elements* alter the wording of the first postulate to make the uniqueness of the line explicit.

Neutral Results in Plane Geometry: TP 6–10

> "... from those propositions of Euclid's first Book that precede the twenty-ninth, wherein begins the use of the disputed postulate...."
>
> —Gerolamo Saccheri, *Euclides Vindicatus.*

The dubious beginning is over. From now on, Lobachevski will deal only with genuine theorems. The important point to observe in these preliminary theorems is that they are *neutral* results: their truth does not depend on Euclid's parallel postulate. Since Euclid delayed his own first use of the parallel postulate until his 29th proposition, Lobachevski is free to use the first propositions I.1–I.28 of the *Elements*.

6) *Vertical angles, those for which the sides of one angle are the extensions of the other, are equal. This is true regardless of whether the vertical angles lie in the plane or on the surface of a sphere.*

Euclid's simple proof of this result (*Elements*, I.15), works on the sphere as well as on the plane.

7) *Two straight lines cannot intersect if a third line cuts them at equal angles.*

That is, if two lines are equally inclined toward a third, then the first two lines will never meet. Euclid proves this in I.28. Since he defines parallels as lines in the same plane that do not intersect one another, Euclid would describe the two lines in this proposition as being parallel. Lobachevski does not use this terminology here; he simply says that the two lines do not intersect one another. We shall see the reason for this in TP 16, where Lobachevski proposes a new definition of the word *parallel*, which these two lines will not satisfy.

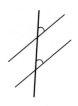

As mentioned above, TP 4 is a special case of TP 7, when the two lines each meet the third at right angles.

8) *In a rectilinear triangle, equal sides lie opposite equal angles, and conversely.*

This is the famous *pons asinorum* (the base angles of an isosceles triangle are equal) and its converse. For proofs, see Euclid (*Elements*, I.5 and I.6).

9) *In rectilinear triangles, greater sides and angles lie opposite one another. In a right triangle, the hypotenuse is greater than either leg, and the two angles adjacent to it are acute.*

This all follows from propositions I.17–I.19 in Euclid's *Elements*:
Euclid proves that in any triangle, the greater of two sides will have the larger opposite angle (I.19), and conversely, the greater of two angles will have the larger opposite side (I.18).

Because Euclid shows that any two angles in a rectilinear triangle sum to less than two right angles (I.17), it follows that in any right triangle, the right angle will be the largest of all of its angles. Consequently, the side opposite the right angle—the hypotenuse—must be the right triangle's largest side (I.18).

10) *Rectilinear triangles are congruent if they have a side and two angles equal, two sides and their included angle equal, two sides and the angle that lies opposite the greatest side equal, or three sides equal.*

This list of triangle congruence criteria includes the familiar four, SAS (*Elements*, I.4), SSS (I.8), ASA (I.26), and AAS (I.26). In general, ASS is not a valid criterion, although it does imply congruence when the angle lies opposite the larger of the two sides. In *The Theory of Parallels*, Lobachevski requires only one sub-case of this criterion: the case in which the angle is *right* (and thus lies opposite the hypotenuse, the largest side). Since Euclid does not prove this "RASS criterion" (*right* angle—side—side), I shall do so to justify Lobachevski's later use of it.

Claim. (RASS) If two right triangles have a leg equal and their hypotenuses equal, then the triangles are congruent.

Proof. Let $\triangle ABC$ and $\triangle A'B'C'$ be the right triangles (with right angles at C and C'), where $AC = A'C'$ and $AB = A'B'$.

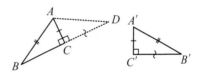

Extend BC to D so that $CD = B'C'$.
Then $\triangle ADC \cong \triangle A'B'C'$ by SAS (*Elements*, I.4).
Hence, $AD = A'B' = AB$, so $\triangle ABD$ is isosceles.
Thus, the base angles at B and D are equal (I.5).
Consequently, $\triangle ABC \cong \triangle ADC$ by AAS (I.26).

Having shown that $\triangle ABC$ *and* $\triangle A'B'C'$ are both congruent to $\triangle ADC$, we conclude that they are congruent to one another, as claimed. ■

Note that the preceding proof does not make use of the parallel postulate. Hence, RASS is a congruence criterion in neutral geometry.

Neutral Results in Solid Geometry: TP 11–15

> "In my hurry I overlooked solid geometry, which should come next, because it's so absurdly undeveloped."
> —Socrates, in Plato's *Republic* (528d)

In several later propositions (TP 26, 27, 34, 35), Lobachevski calls upon a handful of basic neutral theorems of solid geometry, which he has collected in the present section. The reader need not worry about them until reaching those portions of *The Theory of Parallels* that take place in three-dimensional space, at which time he can refer back to this section as needed.

11) *If a straight line is perpendicular to two intersecting lines, but does not lie in their common plane, then it is perpendicular to all straight lines in their common plane that pass through their point of intersection.*

This is one of Euclid's first theorems of solid geometry (*Elements*, XI.4). We say that a line in space is perpendicular to a given plane if it is perpendicular to all lines in the plane that pass through the point at which it pierces the plane. Since there are infinitely many such lines, verifying that a line is perpendicular to a plane could be difficult in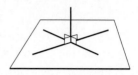
practice were it not for the present theorem. It tells us that once we know that a certain line in space is perpendicular to two lines in a given plane, we may conclude that the line is perpendicular to the plane.

An examination of Euclid's proof shows that this is a neutral result.

12) *The intersection of a sphere with a plane is a circle.*
13) *If a straight line is perpendicular to the intersection of two perpendicular planes and lies in one of them, then it is perpendicular to the other plane.*

I shall prove these two neutral theorems, neither of which appears in Euclid's *Elements*, in reverse order.

To understand TP 13, we must first recall that the angle formed by two planes at their line of intersection is called a *dihedral angle*. We measure a dihedral angle as follows. From an arbitrary point of its "hinge" (the line in which the two planes meet), we erect two perpendiculars, one in each plane. We call these perpendiculars *lines of slope* for the dihedral angle, and we define the dihedral angle's measure to be equal to the measure of the plane angle between the lines of slope. Naturally, we must show that the measure of dihedral angle is a well-defined concept. That is, we must show that it yields the same value no matter which point of the hinge from which we draw the lines of slope. I have given a neutral proof of this fact in the notes to TP 26.[5]

TP 13 concerns perpendicular planes: planes meeting at a dihedral angle of $\pi/2$.

Claim 1 (TP 13). Given a pair of perpendicular planes, if a line lying in one of them makes a right angle with the hinge between them, then that line is perpendicular to the other plane.

Proof. Let α and β be perpendicular planes. Let h be their hinge, their line of intersection.

Let l be a line in α such that $l \perp h$. We must show that $l \perp \beta$.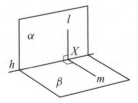

Let X be the point at which l and h meet.

Let m be the line in β that passes through X and is perpendicular to h.

Since l and m are lines of slope for the dihedral angle between the perpendicular planes, the definition of dihedral angle measure tells us that the plane angle between l and m is $\pi/2$. That is, $l \perp m$. Moreover, we already know that $l \perp h$. Hence, l is perpendicular to two lines in plane β, from which it follows that l is perpendicular to plane β (by TP 11).   ∎

---

[5] See the subsection, "A Dihedral Digression", in the TP 26 notes. The proof that the dihedral angle is well defined does not depend on any intermediary work, so the interested reader may examine it immediately.

In future propositions, Lobachevski will often drop a perpendicular from a point to a plane, or erect a perpendicular from a point on a plane. These basic procedures are legitimate in neutral geometry, but the relevant constructions in Euclid's *Elements* (XI.11, 12) involve the parallel postulate. Consequently, we are obliged to legitimize their use in the present context by "neutralizing" Euclid's proofs. That is, we must assure ourselves that we can drop or erect perpendiculars without using the parallel postulate in the process.

Claim 2 (Euclid XI.11—neutralized). Given a plane and a point not on it, we may drop a perpendicular from the point to the plane.

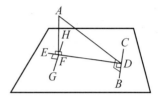

Proof. Let A be the point, and let BC be a random line in the given plane.

In plane ABC, drop a perpendicular AD from A to BC.
In the given plane, erect DE perpendicular to BC at D.
In plane AED, drop a perpendicular AF from A to ED.
We shall show that AF is perpendicular to the given plane.
In the given plane, erect GH perpendicular to ED at F.
Line BC is perpendicular to plane AFD (Euclid XI.4 / TP 11).

Thus, the given plane is perpendicular to plane AFD. (Proof: Erect lines of slope from D, a point on the hinge between the two planes. Since the line of slope in the given plane, DB, is perpendicular to the other plane, it is *a fortiori* perpendicular to the line of slope in it. Since the lines of slope are perpendicular, the dihedral angle between the planes is $\pi/2$, so the planes are perpendicular, as claimed.)

Since GF lies in one of the perpendicular planes (the given plane), and is perpendicular to the hinge between them, it must be perpendicular to the other plane, AFD, by TP 13 (Claim 1).

Thus, AF is perpendicular to GF and DF, both of which lie in the given plane.

Hence, AF is perpendicular to the given plane (Euclid XI.4 / TP 11), as was to be shown. ∎

Claim 3 (Euclid XI.12—neutralized). Given a plane and a point on it, we may erect a perpendicular from the point to the plane.

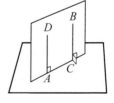

Proof. Let A be the given point on the given plane.
Let B be a random point not on the plane.
Drop a perpendicular BC from B to the plane (by Claim 2).

Plane ABC is perpendicular to the given plane. (Draw lines of slope from C, to a point on the hinge between the two planes. Since BC, which is the line of slope in plane ABC, is perpendicular to the given plane, it is *a fortiori* perpendicular to the line of slope that lies in it. Since these lines of slope meet at right angles, the dihedral angle between the planes is $\pi/2$, so the planes are perpendicular, as claimed.)

In plane ABC, erect a perpendicular AD to line AC at point A.

By TP 13, AD is a line perpendicular to the given plane, which was to be constructed. ∎

Now that we know that we may drop or erect perpendiculars as we please in solid neutral geometry, we shall return to TP 12, which we have yet to prove.

Claim 4 (TP 12). The intersection of a sphere and a plane is a circle.

Proof. To prove that the intersection is a circle, we must show that it consists of all points in the plane that lie at some fixed distance from a particular center. If the cutting plane happens to pass through the sphere's center O, then it is easy to see that the intersection can be characterized as the set of points in the cutting plane whose distance from O is equal

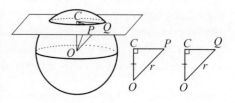

to the radius of the sphere. That is, the intersection is a circle, as claimed.

If the cutting plane does *not* contain O, we proceed as follows.

Drop a perpendicular OC to the plane (Claim 2).

Let P and Q be arbitrary points of the intersection.

We know that $\angle OCP = \angle OCQ = \pi/2$, since OC is perpendicular to the plane. We also know that $OP = OQ$, since P and Q lie on the sphere. Thus, $\triangle OCP \cong \triangle OCQ$, by RASS (TP 10). From this it follows that $CP = CQ$. That is, all points of the intersection are equidistant from C. Calling this common distance r, we have shown that all points of the sphere-plane intersection lie on the circle of radius r whose center is C. It remains only to establish the converse–that every point on this circle is a point of the sphere-plane intersection.

To this end, let X be any point of the circle (and hence of the plane). Then $CX = CP = r$, so $\triangle OCX \cong \triangle OCP$ by SAS. Hence, $OX = OP$, which means that X lies on the sphere. That is, X is part of the sphere-plane intersection. Consequently, the sphere-plane intersection and the circle are identical, as claimed.   ∎

14) *In a spherical triangle, equal angles lie opposite equal sides, and conversely.*

On a sphere, any two non-antipodal points can be joined by a unique great circle. The two points split "their" great circle into a pair of arcs. We form a *spherical triangle* as follows: pick three points on a sphere (no two of which are antipodal), and connect each pair by the shorter of the two great circle arcs that join them.

TP 14 is the spherical analog for Euclid's 5th and 6th propositions (or TP 8, for that matter).

Euclid's proofs for these theorems are in fact valid on the sphere, although we must be careful: at one point in his proof of I.5, he uses his 2nd postulate (a line segment can be extended indefinitely), which does not hold on the sphere. Luckily, when Euclid extends the side of a triangle in this proof, it does not matter how small the extension is. Hence, we can accommodate the extension on the sphere: we always have a little room to extend the sides of a spherical triangle, since they are always strictly less than half the circumference of a great circle.

15) *Spherical triangles are congruent if they have two sides and their included angle equal, or one side and its adjacent angles equal.*

Although Lobachevski mentions only SAS and ASA here, he uses AAS as well in his proof of TP 27. This is not a problem since spherical triangles (when defined as in the notes to TP 14) admit all the congruence criteria that hold for plane triangles, plus one additional one: AAA.

Explanations and proofs shall accompany the theorems from now on.

The preliminary material is over. We shall now enter *The Theory of Parallels* proper.

Theory of Parallels 16

In a plane, all lines that emanate from a point can be partitioned into two classes with respect to a given line in the same plane; namely, those that cut the given line and those that do not cut it.

The boundary-line separating the classes from one another shall be called a *parallel* to the given line.

Lobachevski commences his *Theory of Parallels* by redefining parallelism. This is no mere preliminary matter, but a bold decision to alter a definition that had stood largely unquestioned since ancient times. For a first-time reader, accustomed to the simplicity of Euclid's definition of parallels (coplanar non-intersecting lines), Lobachevski's replacement will no doubt seem mysterious, if not presumptuous. What exactly does it mean? Is it permissible to redefine a familiar term? What is wrong with the classical definition? Why does Lobachevski not simply contrive a new name for his "boundary-line" relation instead of appropriating the term "parallelism"?

We shall answer all of these questions shortly. For now, let us read Lobachevski's description of the geometric configuration that inspired his definition: a configuration directly related to Euclid's parallel postulate.

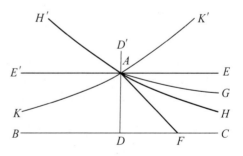

From point A (see Fig. 1), drop the perpendicular AD to the line BC, and erect the perpendicular AE upon it. Now, either all of the lines entering the right angle $\angle EAD$ through A will, like AF in the figure, cut DC, or some of these lines will not cut DC, resembling the perpendicular AE in this respect. The uncertainty as to whether the perpendicular AE is the only line that fails to cut DC requires us to suppose it possible that there are still other lines, such as AG, which do not cut, no matter how far they are extended.

Radical Caution

Consider the various rays that one can draw from point A. Let us call such a ray a "cutting ray" if it intersects ray DC; a "non-cutting ray" if it does not. Since Euclid I.28 (a neutral theorem) implies that AE is a non-cutting ray, it follows that all rays entering $\angle D'AE$ through A are non-cutting as well. (Proof: To cut DC, such a ray would have to cross AE a second time, which is impossible by TP 2.) Of the rays entering $\angle DAE$, it is clear that some will cut DC. Will *all* of them cut DC?

Euclid's parallel postulate asserts that they will.

Yet suppose we ignore the postulate for a moment, and consider the question afresh, in all innocence of geometric tradition. If we choose point X such that $\angle DAX = 89.99999999°$, the human eye cannot distinguish AX from AE. Thus, the evidence of our senses suggests that these two rays will behave similarly: that is, AX "ought" to be a *non*-cutting ray, like its indistinguishable twin, AE.

"Nonsense!" cries a naysayer, "AX is *obviously* a cutting ray. Its approach toward DC is so slow that their intersection might not occur in this galaxy, but nonetheless, it *is* approaching it, so the rays will *eventually* meet."

So it may seem, but is it prudent to presume knowledge of how lines behave over distances so vast that they dwarf all human experience?

Perhaps it is, at least in certain cases. Is this such a case?

The point here is not that one side is right or wrong, but rather that *there is room for debate*. Since a decisive argument would entail proving (or disproving) the parallel postulate, the debate must continue unresolved *ad infinitum*, until one side, in exasperation, ends it at last by formally adopting their own opinion as an article of faith (a postulate), thus rendering further debate impossible.[1] This was Euclid's course. Lest I be misunderstood, let me emphasize that Euclid's assumption of the parallel postulate should not be taken as a sign of argumentative weakness. Considering the number of mathematicians throughout two millennia who believed that the parallel postulate could be proved, Euclid's insightful recognition that it must be assumed stands as testimony to his genius. The late 19th-century proof that Euclid's fifth postulate is *not* a logical consequence of his first four vindicated not only Lobachevski, but Euclid as well.

Lobachevski exercised radical caution and restraint with respect to the parallel postulate. He acknowledged that *we do not know* whether it holds in physical space; he suggested that we may never know, except perhaps through the analysis of future astronomical measurements; and consequently, since there is room for debate, he believed that we should not presume the answer. Euclid might, after all, have been wrong. Instead, we should examine both possibilities, thus preparing ourselves for either eventuality, should we ever learn the truth.

Accordingly, Lobachevski asks us to consider the *possibility* that a ray may exist, which enters $\angle DAE$ but fails to intersect DC. By tracing the consequences that would follow, he developed the first non-Euclidean geometry. We begin this long journey with a simple observation. If there is one such ray, then there will be infinitely many: for if AG is a non-cutting ray, all rays entering $\angle EAG$ must also fail to cut DC. We are thus confronted with a picture of two segregated groups of rays. Lobachevski will describe this picture next, but rather than cutting (or non-cutting) *rays*, he refers to cutting (or non-cutting) *lines*. These are simply the lines that contain the rays in question.

---

[1] Cf. Bertrand Russell's description of his first encounter with Euclid at age eleven: "I had been told that Euclid proved things, and was much disappointed that he started with axioms. At first I refused to accept them unless my brother could offer me some reason for doing so, but he said: 'If you don't accept them we cannot go on', and as I wished to go on, I reluctantly admitted them *pro tem*." (Russell, *Autobiography*, p. 38.)

At the transition from the cutting lines such as AF to the non-cutting lines such as AG, one necessarily encounters a parallel to DC. That is, one will encounter a boundary line AH with the property that all the lines on one side of it, such as AG, do not cut DC, while all the lines on the other side of it, such as AF, do cut DC.

There will be one ray that acts as the boundary between those that cut and those that do not. The boundary ray is a non-cutting ray[2], but in contrast to the other non-cutting rays, this one admits no "wiggle room": if we rotate the boundary ray about A towards ray DC, then regardless of how minuscule the rotation, it will always intersect DC, since *every* ray below the boundary is, by definition, a cutting ray.

According to Lobachevski's definition, the boundary ray is parallel to DC. (More accurately, he defines the line containing the boundary ray to be parallel to DC.) We shall now examine his definition, and consider his reasons for adopting it.

A Deeper Definition

Let AB and CD be two coplanar lines. Euclid calls them parallel if and only if they do not meet. Lobachevski, however, insists that parallels should satisfy a second condition as well.

Lobachevski's Definition of Parallelism

If AH and DC are coplanar lines, then AH is *parallel* to DC (in symbols, $AH\|DC$) if:

1) The lines do not meet, *and*
2) There is no "wiggle room".
   (That is, every ray AX that enters $\angle HAD$ intersects DC.)

Important Note: $AH\|DC$ is *not* equivalent to $HA\|CD$. In the former case, line DC is cut by all rays entering $\angle HAD$. In the latter, the same line is cut by all rays entering $\angle AHC$. It is not hard to show that if one of these conditions holds, the other necessarily fails.

This definition takes some time to digest, and it invariably raises questions, some of which Lobachevski does not answer until TP 25, some of which he does not even address at all. I shall anticipate some of these questions, provide their answers, and indicate where Lobachevski gives his own answers.

1. *Lobachevski's words in TP 16 suggest that D should be the foot of a perpendicular dropped from A to the second line. Why is this not part of the definition?*

I have deliberately omitted it because it is distracting and irrelevant. It is a trivial exercise to show that if $AH\|DC$, then $AH\|XC$ for *any* choice of X on the second line, DC.

---

[2] *Proof.* Suppose, by way of contradiction, that the boundary-ray AH is a cutting ray. Then it cuts DC at some point X. Choose any point $Y \in DC$ to the right of X, and draw the line AY. Since AY lies above the boundary-line AH, it must be a non-cutting ray, by definition of the boundary-line. However, by its very construction, we know that AY cuts DC at K. Contradiction.

2. *Parallelism is a relation between two lines, but* point A *seems to play a very special role in Lobachevski's definition. This poses a problem: if* $AH\|DC$, *and* $P \in AH$, *we should expect that* $PH\|DC$, *since* PH *and* AH *are just different names for the same line. Is this actually the case?*

Yes, it is. (The proof is in TP 17.)

3. *Parallelism should be symmetric:* $AH\|DC$ *should imply* $DC\|AH$. *This is obvious under Euclid's old definition of parallelism. Is it still true under Lobachevski's new definition?*

Yes, it is. (The proof is in TP 18.)

4. *Parallelism should be transitive: if two lines are parallel to a third, then they should also be parallel to each other. Does this follow from Lobachevski's definition?*

Yes, it does. (The proof is in TP 25.)

5. *Lobachevski is supposed to be addressing problems in geometry that have plagued mathematicians for millennia; by redefining a key term, isn't he actually avoiding the old problems, rather than confronting them?*

No, he is not; Lobachevski's definition is a generalization of Euclid's own. That is, if we were to accept the parallel postulate, the two definitions would be logically equivalent[3]. Therefore, a traditional geometer, one content to accept the parallel postulate, can raise no logical objections to Lobachevski's definition. One may reasonably complain that it is unwieldy or uneconomical, but so long as we retain the parallel postulate, it is just a different way of saying the same thing; using it is as harmless as doing geometry in Spanish rather than English: the words have changed, but the theorems and problems remain the same.

6. *In that case, why bother with a new definition? After all, Euclid's is easier to understand.*

For one who does not wish to question the parallel postulate, there is no need to bother. But for one who adopts Lobachevski's more cautious stance, the new definition will prove itself considerably more robust. If we accept the parallel postulate, the two definitions are equivalent, but in the wider context of neutral geometry (where we assume only the first four postulates), Euclid's definition reveals its weakness: it is not transitive. For example, Euclid would say that AE and AG (in Lobachevski's figure) are both parallel to DC, but he would have to admit that they are not parallel to one another since they meet at A. In contrast, Lobachevski's definition of parallelism retains its transitivity, even if we do not accept the parallel postulate. (See TP 25.)

To recapitulate—in Euclidean geometry, the two definitions are interchangeable; in the larger context of neutral geometry, Lobachevski's definition is superior. It is therefore a *deeper* definition: it incorporates the old definition as a special case, while successfully extending the notion of parallelism to a broader setting. We may consider ourselves fully justified in using it. One last question, a subtle one, which I have hitherto sidestepped, remains.

---

[3] *Proof.* Since lines satisfying Lobachevski's definition do not intersect, they obviously satisfy Euclid's definition. Conversely, if AH and DC satisfy Euclid's definition, then $\angle HAD + \angle CDA = 180°$ (by Euclid I.29); this being the case, the parallel postulate itself rules out the possibility of "wiggle room", so we may conclude that AH and DC satisfy Lobachevski's definition as well.

7. Must a parallel exist in Lobachevski's figure? In other words, must there be a "last" non-cutting line? Or might the non-cutting lines resemble the positive real numbers, being bounded below, but without a least member?

We can secure the existence of our parallel as follows.

Every ray AX entering the angle $\angle EAD$ corresponds to a real number between 0 and $\pi/2$ (the radian measure of the angle $\angle XAD$) and conversely. The set of real numbers corresponding to the cutting rays is bounded above by $\pi/2$, since this number corresponds to AE, a ray known to be non-cutting. Any bounded set of reals has a least upper bound, and the ray corresponding to this least upper bound is easily seen to be the boundary between the cutting and non-cutting lines. Thus, the parallel exists.

Homogeneity and the Angle of Parallelism

The angle $\angle HAD$ between the parallel AH and the perpendicular AD is called the *angle of parallelism*; we shall denote it here by $\Pi(p)$, where $p = AD$.

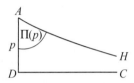

Given any line segment (for example, AD in the figure), if we draw two rays from its endpoints such that the first ray is perpendicular to the segment and the second ray is parallel to the first, then the angle between the second ray and the segment is the segment's "angle of parallelism".

Like Euclid, Lobachevski implicitly assumes the *homogeneity* of space. That is, he assumes that empty space "looks the same" from every point, and from every direction[4]. The plane has no crinkles or other irregularities. An important consequence of this assumption is that if we take two line segments of the same length to distinct locations, and then carry out the same set of constructions upon each of them, the homogeneity of space ensures that the resulting figures are congruent to one another.

Consequently, line segments of the same length will have the same angle of parallelism, which explains why Lobachevski expresses the angle of parallelism as a function of length, p.

The Path to Imaginary Geometry

If $\Pi(p)$ is a right angle, then the extension AE' of AE will be parallel to the extension DB of the line DC. Observing the four right angles formed at point A by the perpendiculars AE, AD, and their extensions AE' and AD', we note that any line emanating from A has the property that either it or its extension lies in one of the two right angles facing BC. Consequently, with the exception of the parallel EE', all lines through A will cut the line BC when sufficiently extended.

Using the ideas in this passage, we can easily demonstrate an important theorem.

---

[4] Obvious expressions of homogeneity in Euclid's *Elements* include the fourth postulate ("all right angles are equal") and the use of superposition in the proofs of I.4 and I.8 (the SAS and SSS congruence criteria).

Claim 1. The parallel postulate holds if and only if $\Pi(p) = \pi/2$ for *all* lengths p.

Proof. $\Rightarrow$) If the postulate holds, then $\Pi(p)$ is obviously a right angle for all lengths p.

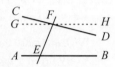

$\Leftarrow$) If $\Pi(p) = \pi/2$ for *all* lengths p, then Lobachevski's words in the preceding passage tell us the following: *if l is any line and P any point not on it, then there is exactly one line through P that does not intersect l*. This statement, commonly known as Playfair's axiom, implies the parallel postulate, as we shall now demonstrate.

Suppose that two lines, AB and CD, cut by a transversal EF, as in the figure, such that $\angle FEB + \angle EFD < 180°$. Draw the unique line GH through E that makes $\angle FEH + \angle EFD = 180°$. By Euclid I.28, GH will not intersect CD. By Playfair's axiom, every other line through E will intersect CD. In particular, EB will intersect CD. Hence, the parallel postulate holds, as claimed. ∎

We have suspended judgment on the parallel postulate, but we are approaching a fork in the road, where we must choose one path or the other. On one path, $\Pi(p)$ is always a right angle; on the other, it must sometimes be acute (In fact, we shall see that it will always be acute on the second path). Lobachevski's policy of radical caution in geometry dictates that we must explore both paths; after all, either one could turn out to be the geometry of physical space. For over 2000 years, Euclid's parallel postulate had acted as a barricade, directing all traffic toward the first path, which leads to Euclidean geometry (which Lobachevski called *the ordinary geometry*). In contrast, the second road leads to unexplored territory. Since no one had ever mapped it out (apart from some brief sketches made by unwitting trespassers, such as Saccheri and Lambert), Lobachevski devoted his own energies to the task. This second road leads to what he called *imaginary geometry*, a term that he will first use at the end of TP 22.

Parallelism has Direction in Imaginary Geometry

If $\Pi(p) < \pi/2$, then the line AK, which lies on the other side of AD and makes the same angle $\angle DAK = \Pi(p)$ with it, will be parallel to the extension DB of the line DC. Hence, under this hypothesis we must distinguish directions of parallelism.

This follows from homogeneity of the plane.

Let AD be the perpendicular dropped from a point A to a line l. In general, there will be a line AH that is parallel to l "towards the right" and a second line AK parallel to l "towards the left". If $\Pi(AD)$ is a right angle, then these two parallels will coincide, but if the angle of parallelism is acute, then the parallels will be distinct.

The existence of a second parallel might seem to contradict my claim that Lobachevski's definition of parallelism is transitive: there are now two lines parallel to l, and they cannot possibly be parallel to one another, since they intersect at A. This problem vanishes if we associate a direction with parallelism, in which case transitivity means that two lines which are parallel to a third *in the same direction* will be parallel to one another in the same direction. Consequently, the failure of AK and AH to be parallel is not a violation of transitivity after all: these two lines are parallel to l in opposite directions. It is for this reason that Lobachevski is careful to distinguish that while AH is parallel to DC (i.e. AH is parallel to the line l in the direction indicated by the ray DC), AK is parallel to DB (i.e. AK is parallel to the line l in the opposite direction as indicated by the ray DB).

Lobachevski's Summary

Among the other lines that enter either of the two right angles facing BC, those lying between the parallels (i.e. those within the angle $\angle HAK = 2\Pi(p)$) belong to the class of cutting-lines. On the other hand, those that lie between either of the parallels and EE' (i.e. those within either of the two angles $\angle EAH = \pi/2 - \Pi(p)$ or $\angle E'AK = \pi/2 - \Pi(p)$) belong, like AG, to the class of non-cutting lines.

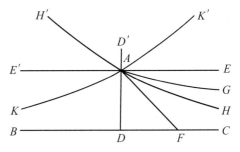

Similarly, on the other side of the line EE', the extensions AH' and AK' of AH and AK are parallel to BC; the others are cutting-lines if they lie in the angle $\angle K'AH'$, but are non-cutting lines if they lie in either of the angles $\angle K'AE$ or $\angle H'AE'$.

Consequently, under the presupposition that $\Pi(p) = \pi/2$, lines can only be cutting-lines or parallels. However, if one assumes that $\Pi(p) < \pi/2$, then one must admit two parallels, one on each side. Furthermore, among the remaining lines, one must distinguish between those that cut and those that do not cut. Under either assumption, the distinguishing mark of parallelism

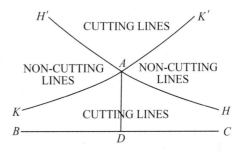

is that the line becomes a cutting line when subjected to the smallest deviation toward the side where the parallel lies. Thus, if AH is parallel to DC, then regardless of how small the angle $\angle HAF$ may be, the line AF will cut DC.

Having demonstrated that an acute angle of parallelism implies a *pair* of parallels through the point A, Lobachevski concludes TP 16 by reiterating the behavior of the non-parallel lines through A, and summarizing the results he has obtained so far. In particular, he draws attention to the fact that when the angle of parallelism is acute, the lines passing through A fall into three classes. Namely, those intersecting BC, those parallel to BC, and those that neither intersect BC nor are parallel to it. Lines of the third type, which do not exist in Euclidean geometry, are sometimes called *ultraparallels* or *hyperparallels* by modern authors[5].

---

[5] A potentially misleading designation. The prefixes "ultra" and "hyper" refer in this context only to the fact that such lines lie "above" the parallels. Words like hypersensitive or ultraconservative might lead one to suspect that ultraparallels possess all the ordinary characteristics of parallels and then some. This is not the case. To cite one obvious example, 'ultraparallelism' is not a transitive relation.

Problematic Pictures

In geometry, a triangle's sides are supposed to be perfectly straight, and devoid of thickness. Triangles drawn with pen and paper can approximate this perfection, but they cannot actually attain it. In practice, this disturbs no one, for we recognize that our pictures are simply representations that assist our reasoning about the real triangles (those inhabiting Plato's world of ideal forms, as it were).

However, imaginary geometry poses additional problems of representation, which do trouble beginners. In TP 23, we shall see that imaginary geometry admits line segments with arbitrarily small angles of parallelism. Suppose that we wish to depict a segment whose angle of parallelism is 30°, together with rays emanating from its endpoints, the first of which is perpendicular to the segment, the second of which is parallel to the first. If we draw this in such a way as to represent the angles accurately, we will immediately run into a dilemma: the rays, which ought to be parallel, will clearly meet.

One compromise is to draw the rays very short, so that their intersection will not actually be depicted on the page. Unfortunately, if we adopt this strategy, we must constantly remind ourselves, "these lines are supposed to represent parallels, even though they don't look parallel at all in the figure."

A second possible compromise, one that I often employ, preserves the appearance of non-intersection by sacrificing the appearance of straightness. That is, we draw the second ray as a curve asymptotic to the first ray. This forces us to bear in mind, "this is a representation of a straight line, although it doesn't look straight on the page." Moreover, if we represent one line by a curve and the other by a straight segment (in an attempt to minimize our infelicities), our representation of the plane will immediately appear to have a favored direction—a direction in which "straight lines look straight". Thus, our plane's representation will not look homogeneous, despite the fact that it is homogeneous in reality. We must constantly remember, "there is nothing special about this direction, despite the drawing."

The moral of the story is that in imaginary geometry, the relationship between a geometric figure and its representation on the page is more complex than it is Euclidean geometry. Provided one keeps this in mind, one quickly becomes accustomed to distorted representations, and learns to read them comfortably. Do not let such representations mislead you into thinking that parallels are not straight in imaginary geometry, or that the plane is not homogeneous. The one sacred relation that we shall always depict accurately is intersection (or lack thereof): lines that do (or do not) intersect one another will faithfully appear that way on the page. It is to preserve this appearance that we sacrifice others.

A skeptic might claim, "These very problems of representation indicate that imaginary geometry is utter nonsense: they arise precisely because the parallel postulate actually *does* hold in the physical universe, the space which also includes our paper and pencils!" This is a thought-provoking claim, but imaginary geometry cannot be disposed of so easily. We shall soon learn (in TP 23) that although imaginary space is homogeneous (it looks the same at every point), it looks very different at different *scales*. On a tiny scale, it resembles Euclidean geometry, and serious deviations become noticeable only on a large, possibly astronomical, scale. Since similar figures do not exist in imaginary geometry (see the notes to TP 20), accurate scaled down drawings are impossible. Thus, if a line segment with a 30° angle of parallelism is several light-years long, then even if our universe is governed by *imaginary* geometry, we have no way to depict it accurately as a three-inch drawing. We would need a piece of paper that would cover much of the galaxy.

A Digression on Rigor in Geometry

In the section labeled "The Path to Imaginary Geometry", we examined a passage in which Lobachevski considers a pair of perpendicular lines that cross at A. These lines (which I shall call the horizontal and vertical axes) naturally divide the plane into four quadrants. Lobachevski asserts that any line through A necessarily enters one of the two lower quadrants below the horizontal axis. This seems so obvious as to render commentary unnecessary, but I wish to dwell upon it for a moment since it will serve well to illustrate a profound philosophical shift that overtook the discipline of mathematics within fifty years of Lobachevski's death.

For Lobachevski's assertion to be false, a line would have to lie entirely above the horizontal axis, touching it once at A, but never actually cutting through to the other side. In other words, the line would have to be tangent to the horizontal axis. Both Euclid and Lobachevski would have considered this situation (two straight lines, tangent to one another at a point) so obviously absurd that a proof of its impossibility would be superfluous. David Hilbert, in contrast, might have proposed a smug inversion: the only thing obvious about Lobachevski's assertion is that if either he or Euclid had tried to prove it rigorously, they would have found the task impossible.[6]

Both attitudes are reasonable, according to their own philosophies.

Let us begin with Hilbert's criticism. Hilbert might point out that if the foundations upon which Euclid based geometry seem capable of supporting such majestic mathematics as the Pythagorean Theorem and the theory of regular polyhedra, it is only because they have been unconsciously wedded to intuitive yet logically unjustified ideas about how lines ought to behave. When examined in the hypercritical mindset that demands rigorous demonstrations of even the most obvious assertions, one finds that the *Elements*, the book venerated for millennia as the pinnacle of human reasoning, is in fact riddled with logical lacunae. Infamously, one discovers that these gaps begin in the very first proposition, and continue to accumulate throughout the thirteen books. Viewed in this harsh unforgiving light, Euclid's masterwork resembles a stately yet dangerous old mansion whose architecture suggests eternal soundness even as its unseen foundations threaten to crumble away.

To prove Lobachevski's assertion that any line passing through A must actually *cut through* the horizontal axis, one must construct a purely logical argument, every statement of which is grounded in the axioms (no appeals to pictures or intuition!), demonstrating that on any line through A, there exists a pair of points separated by the horizontal axis. That is, one must be able to distinguish between the two regions of the plane lying to either side of a line.

Hilbert can accomplish this with the "betweenness" axioms[7] that he built into his foundations of geometry. "Betweenness" is one of Hilbert's undefined concepts, and his axioms concerning this relation (for example, "if A and C are two points on a line, there exists at least one other point B on the line that lies between A and C.") yield theorems capable of distinguishing between

---

[6] I am using Hilbert's name here to represent the work of all those mathematicians who worked on the foundations of geometry in the late 19th and early 20th centuries.

[7] These derive from the work of Moritz Pasch, whose *Vorlesungen über neuere Geometrie* (1882) was the first major work devoted to reinforcing the foundations of geometry.

the two regions into which a line divides a plane; the interior and exterior of a triangle; and other fundamental concepts that Euclid leaves entirely to intuition. Since these axioms have no Euclidean analogue (Robin Hartshorne has accordingly called them "the most striking innovation in this set"[8]), any attempt to prove a statement about passing from one region to the other in Euclid's system is doomed to failure. Hilbert's criticism would seem therefore to be valid.

This raises an intriguing question. When Lobachevski made assertions such as the one singled out above, didn't he realize that he was relying upon instinct rather than logic? One would think that he, of all people, would have been acutely sensitive to such issues; his work in non-Euclidean geometry is itself a profound and extended meditation upon an axiom.

In fact, Lobachevski contemplated foundational issues deeply and broadly. In his largest work, *New Principles of Geometry with a Complete Theory of Parallels* (1835–8), for example, he endeavored to base all of geometry on the topological concepts of touching and cutting. In this vein, he proposed that solid bodies, rather than points, are the true fundamental geometric entities; surfaces, for example, should be understood in terms of solids, of which they are abstractions; curves arise as sections of surfaces and so forth. Leibniz, incidentally, was of the same opinion. Both Hilbert and Lobachevski devoted much attention to foundations, but motivated by distinct philosophies, they followed quite different paths of inquiry.

For Lobachevski, and Lobachevski's age, geometry was still the study of forms occurring in the physical universe, or abstractions thereof. According to this view, the axioms of geometry must be basic self-evident truths. In turn, theorems deduced from these axioms reveal aspects of physical reality not directly evident to the senses, yet irrefutable all the same, since they derive wholly from immaculate sources—a set of self-evident axioms and the pure methods of logical deduction. Naturally, when one conceives of geometry this way, the proper choice of axioms is crucial. An axiom asserting a statement contrary to the true nature of space would seriously compromise geometry's accuracy as physical description.

Euclid's parallel postulate, Lobachevski suspected, might be such an axiom. As discussed earlier, one cannot verify it in a physical setting. Imagine two drawings: one of a transversal cutting a pair of lines so that the angles it makes with them on one side add up to 179.999999°, the other drawing depicting the same situation except that the angles add up to 180°. The human eye cannot distinguish such small differences. We can prove (via the first four of Euclid's postulates) that the lines in the second drawing will never meet. If the lines of the first drawing, indistinguishable from those of the second, somehow manage to intersect, this should come as a great surprise to us, as an instance of physical reality contradicting the evidence of the senses. Such surprises can be delightful when deduced as theorems, but as axioms, they are dubious indeed.

Lobachevski never claimed that the parallel postulate does not hold in reality, only that we do not know for certain whether it does. The difficulty stems from the fact that the parallel postulate makes assertions about intersections that occur at indefinite, possibly unfathomable distances. Even with eyesight sufficiently sharp to distinguish between the two drawings in the thought experiment above and even to observe the lines in the first drawing coming closer together, we would remain unable to verify an eventual intersection, which, if it does occur, might happen millions of light-years away. Bound as we are to an insubstantially small portion of the universe,

---

[8] Hartshorne, p. 65

such large-scale phenomena defeat our powers of observation, and therefore we must exercise caution in making statements about them.

By way of contrast, let us reconsider the assertion that intersecting lines must actually cross one another. This statement concerns only the small-scale, local behavior of lines in the immediate vicinity of a specific point. Indefinite distances are not involved. Our physical experience with straight lines, whether drawn by hand or occurring in nature, leaves us with no doubt that this assertion is true. Consequently, Lobachevski's omission of a proof is, according to his philosophy, of trifling significance. Had someone specifically questioned him about this, he almost certainly would have acknowledged the existence of a gap in the structure of his argument. He would have been justified all the same in dismissing this gap as innocuous and his questioner as pedantic. Lines, in Lobachevski's mind, were forms with definite intrinsic properties. If the axioms of geometry failed to capture those properties, the fault lay with the axioms. An obvious fact that eludes proof remains a fact nonetheless.

While Lobachevski's work drew inspiration from the relationship of mathematics to the natural world, a major impetus for rebuilding the foundations in the years around 1900 was Lobachevski's own discovery, non-Euclidean geometry. Mathematicians were forced to proceed with unusual care while learning or developing this new subject, exercising caution not to inadvertently use a theorem that relied upon the parallel postulate. With their critical attention heightened out of necessity, many mathematicians began to pay more attention to the little holes in the foundations that they had previously ignored, and they noted with some alarm the existence of more holes than they had suspected. The efforts to repair, if not rebuild, the foundations began shortly thereafter. The mathematicians who led these efforts emphasized that, if geometry is to be a truly deductive science, one must be able to trace any geometric theorem back to the axioms using logic alone. Intuition may be useful as a guide, but it may never substitute for logic. Physical reality might suggest the axioms, but once they are decided upon, appeals to physical forms are inadmissible in a rigorous proof. Hilbert insisted that the lines of geometry have no platonic existence, and thus have no intrinsic properties other than those with which the axioms endowed them.

Consequently, many of Lobachevski's arguments cannot stand up to Hilbertian criticism. This does not render them invalid. Indeed, not one of his theorems collapses under the strain of the newly imposed rigor, even if their proofs can be improved by erecting them over Hilbert's new airtight foundations. For this reason (coupled with the fact that geometric proofs that leave nothing to the imagination tend to be quite long and somewhat repellent), I shall not dwell overmuch upon Hilbert's foundations in these pages except where they convey extra insight into Lobachevski's mathematics.

Violets in Spring

> "...many things have an epoch in which they are found at the same time in several places, just as violets appear on every side in spring."
>
> —Farkas Bolyai.

Lobachevski, Gauss, and Bolyai independently redefined parallelism essentially the same way, at approximately the same time. Gauss confided few of his ideas on non-Euclidean geometry to paper, and made none of them public, but two brief memoranda containing his definition of parallels along with a few relevant theorems (including a proof of the transitivity of parallelism) were found among his personal papers after his death in 1855. These were published posthumously in his complete works[9].

Bolyai defines parallelism in the first sentence of §1 of his *Appendix*. More accurately, Bolyai defines the relation that Lobachevski and Gauss call "parallelism"; surprisingly, he does not actually use the word himself. His definition reads, "If a ray AM is not cut by a ray BN, situated in the same plane, but is cut by every ray BP in the angle $\angle ABN$, this is designated by $BN|||AM$."[10] This definition, which is clearly equivalent to (and, it must be admitted, stated more concisely than) Lobachevski's, was apparently first suggested to Bolyai by Carl Szàsz, a friend from his years (1817–22) at the Royal College for Engineers in Vienna, with whom Bolyai had frequently discussed geometry.[11]

---

[9] Gauss, pp. 202–209. A detailed exposition (in English) of Gauss' notes on the definition of parallelism is in Bonola, pp. 67–75.

[10] Halsted inserts the phrase, "we will call ray BN *parallel* to ray AM" into his translation; it does not occur in Bolyai's original Latin.

[11] Gray, *János Bolyai*, p. 50.

Theory of Parallels 17

A straight line retains the distinguishing mark of parallelism at all its points.

In TP 17, Lobachevski proves that his new sense of parallelism[1] is well defined.

Recall that $AB||CD$ only if AB admits no wiggle room about A. (i.e. AB must exhibit the "mark of parallelism" at A.) Since A has no particular significance among the infinitely many points on line AB, its conspicuous presence in the definition of parallelism is disconcerting. To set our minds at ease, Lobachevski demonstrates that A's ostensibly special role is an illusion: he proves that if the line exhibits the mark of parallelism (lack of wiggle room) at any one of its points, then it will exhibit the mark at *all* of its points. Therefore, parallelism does *not* depend upon any particular point.

Let AB be parallel to CD, with AC perpendicular to the latter. We shall examine two points, one chosen arbitrarily from the line AB and one chosen arbitrarily from its extension beyond the perpendicular.

Lobachevski's proof requires two cases. Assuming that $AB||CD$, he first shows that $EB||CD$, where E is an arbitrary point of ray AB; he then shows that $E'B||CD$, where E' is a point chosen arbitrarily from the rest of line AB.

First Case (and Interlude with Pasch)

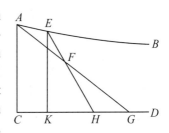

Let E be a point on that side of the perpendicular in which AB is parallel to BC. From E, drop a perpendicular EK to CD, and draw any line EF lying within the angle $\angle BEK$. Draw the line through the points A and F. Its extension must intersect CD (by TP 16) at some point G. This produces a triangle $\triangle ACG$, which is pierced by the line EF. This line, by construction, cannot intersect AC; nor can it intersect AG or EK a second time (TP 2). Hence, it must meet CD at some point H (by TP 3).

The idea behind the first case is straightforward, although it may be worth noting that the coup de grace—Lobachevski's claim that EF must exit $\triangle ACG$ through side CG—is another

---

[1] See the section, "A Deeper Definition" in the notes to TP 16.

example of an intuitively obvious statement that cannot be rigorously justified on the basis of Euclid's axioms. Hilbert would justify this claim by appealing to one of his betweenness axioms, known today as *Pasch's axiom*. Pasch's axiom asserts that if a line intersects one side of a triangle, but does not pass through any of its vertices, it will intersect one (and only one) of the other sides as well[2].

Where Hilbert would call on Pasch's axiom, Lobachevski invokes his own TP 3, a very general "what-goes-in-must-come-out" theorem, applicable not only to lines piercing triangles, but also to lines piercing any bounded region. It cannot be proved as a theorem using Euclid's foundations, and thus, strictly speaking, should be considered an axiom in Lobachevski's development of geometry.

Since TP 3 deals with arbitrary bounded regions, as opposed to mere triangles, it may appear to be much stronger than Pasch's axiom. However, in the specific case of triangles, it is actually the weaker of the two, inasmuch as it provides no detail as to *where* the line will exit the triangle. To demonstrate this weakness, we shall review the last two steps of the proof above and compare how Hilbert and Lobachevski would justify them. In doing so, we shall see that TP 3 is not quite powerful enough to establish the intersection of EF and CD.

The steps are as follows: 1. Line EF enters triangle $\triangle ACG$, so it must exit the triangle as well. 2. Since it cannot leave through either side AC or side AG, it must pass through CG. (Lobachevski's comment that it cannot intersect EK a second time is true, but superfluous as far as the proof is concerned.)

Both Hilbert and Lobachevski can justify step one with an axiom (Pasch's axiom or TP3 respectively).

As for step two, Hilbert has the upper hand. Since EF intersects side AG, Pasch's axiom asserts that it must cut either AC or CG on its way out of the triangle. Hilbert can easily demonstrate that EF cannot intersect AC (although this takes a little work when arguing from first principles), and may therefore conclude that it passes through CG as claimed. Lobachevski, on the other hand, runs into several problems. To cite just one example, the possibility that EF might exit the triangle through AG (the same side through which it entered) is not ruled out by TP 3, so Lobachevski must prove this. In an attempt to do so, he invokes TP 2. This almost works, but a tiny hole remains in his demonstration—TP 2 does rule out the possibility that EF might leave AG via the same point through which it entered. Absurd as this possibility sounds, Lobachevski lacks the logical apparatus to expose it as such. If pressed, he might appeal to the enigmatic TP 1, which, as we have argued in the notes to that proposition, amounts to an article of faith that certain behavior, such as self-intersection, is "repugnant to the nature of a straight line" and consequently, need not be considered at all.

Second Case

Now let E' be a point on the extension of AB, and drop a perpendicular $E'K'$ to the extension of the line CD. Draw any line $E'F'$ with the angle $\angle AE'F'$ small enough to cut AC at some

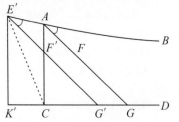

---

[2] From this axiom and some very basic results, one can prove the related *crossbar theorem*: if a line enters the interior of a triangle through one of its vertices, it must exit through the opposite side. Together, Pasch's axiom and the crossbar theorem guarantee that if a line enters the interior of the triangle, it must also leave it.

point F'. At the same angle of inclination towards AB, draw a line AF; its extension will intersect CD (by TP 16) at some point G. This construction produces a triangle $\triangle AGC$, which is pierced by the extension of line $E'F'$. This line can neither cut AC a second time, nor can it cut AG, since $\angle BAG = \angle BE'G'$ (by TP 7). Thus, it must meet CD at some point G'.

The second case involves an argument similar to that used in the first, though slightly more involved. We want to prove that $E'B\|CD$, so according to our definition of parallelism, we must show that all rays entering angle $\angle BE'C$ must intersect CD. All such rays cut AC (apply the crossbar theorem, which was mentioned in the last footnote, to $\triangle AE'C$), so when Lobachevski restricts his attention to those rays that cut AC, there is no loss of generality.

Therefore, regardless of which points E and E' the lines EF and $E'F'$ emanate from, and regardless of how little these lines deviate from AB, they will always cut CD, the line to which AB is parallel.

By demonstrating that the mark of parallelism propagates throughout the entire line once it appears at any one point, TP 17 confirms that parallelism is a relation strictly between lines, without dependence upon an intermediary point.

Bolyai, Gauss, and What Might Have Been

As mentioned in the notes following TP 16, Bolyai's definition of parallelism (§1 of his *Appendix*) is expressed in the language of *rays*: If a ray CD is not cut by a ray AB, but is cut by every ray AP in the angle $\angle BAC$, Bolyai defines AB to be parallel to CD. In §2, he proves that parallelism between rays depends only upon their directions, not their initial points. This is, of course, analogous to Lobachevski's TP 17, and Bolyai's proof is essentially identical to the proof we have just examined.

Gauss also works with rays. His proof is identical to Bolyai's, although his writing is considerably more lucid. Indeed, Gauss' few surviving personal memoranda on non-Euclidean geometry are notable for their clarity of exposition. He had intended to compose, though probably never to publish, a full treatise on non-Euclidean geometry, as a means of ensuring that an account of this subject would survive him. After reading Bolyai's *Appendix*, however, he considered himself released from this burden. The world might be richer today had Gauss never read the works of Bolyai and Lobachevski. We would possess not only a beautiful Gaussian treatise on non-Euclidean geometry, but perhaps more geometric works of an undiscouraged János Bolyai as well.

Theory of Parallels 18

Two parallel lines are always mutually parallel.

Lobachevski is going to show that parallelism is a *symmetric* relation: given $AB \parallel CD$, he will prove that $CD \parallel AB$. To do so, he must verify that every ray CE entering $\angle DCA$ intersects AB. Clever though his proof is, Lobachevski's obscures his geometric artistry under murky exposition. Accordingly, I shall follow his proof with an alternate explanation of my own.

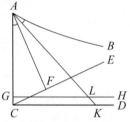

Let AC be perpendicular to CD, a line to which AB is parallel. From C, draw any line CE making an acute angle $\angle ECD$ with CD. From A, drop the perpendicular AF to CE. This produces a right triangle $\triangle ACF$, in which the hypotenuse AC is greater than the side AF (TP 9).

If we make $AG = AF$ and lay AF upon AG, the lines AB and FE will assume positions AK and GH in such a way that $\angle BAK = \angle FAC$. Consequently, AK must intersect the line DC at some point K (TP 16), giving rise to a triangle $\triangle AKC$. The perpendicular GH within this triangle must meet the line AK at some point L (TP 3). Measured along AB from A, the distance AL determines the intersection point of the lines AB and CE.

Therefore, CE will always intersect AB, regardless of how small the angle $\angle ECD$ may be. Hence, CD is parallel to AB (TP 16).

The Idea Behind the Proof

The best way to understand Lobachevski's proof is to imagine that we have two identical (i.e. congruent) copies of the figure $ABCDEF$, the first drawn on an opaque piece of paper, the second on a transparent sheet of plastic. To distinguish the two, we shall put primes on the letters of the second copy. Lay the transparent sheet on top of the opaque one so that the two figures coincide point for point, as in the first figure below.

Keeping the bottom sheet fixed, rotate the top one about A until $A'F'$ lies on AC, as in the second figure. By the definition of parallelism, $A'B'$ must cut CD, ($AB \parallel CD$ and ray $A'B'$ enters $\angle BAC$). Yet in order to cut CD, ray $A'B'$ must first intersect $C'E'$. That is, $C'E'$ and $A'B'$ are intersecting lines in figure $A'B'C'D'E'F'$. Hence, the corresponding lines on the congruent figure $ABCDEF$ also intersect one another. Namely, CE intersects AB, which was to be shown.

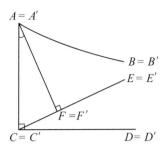
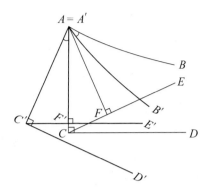

Last Thoughts on TP 18

If one insists on Hilbertian rigor, one must either produce a rigorous definition of rotation and prove that this operation possesses all the properties that we expect of it[1], or one must abandon the ill-defined procedure of "laying upon," and construct a proof more rigorous, though inevitably less direct, than Lobachevski's. Gauss' memoirs contain a proof along such lines; it is clever, meticulous, not particularly transparent, and involves two cases. Bolyai, for his part, establishes the symmetry of parallelism in §5 of the *Appendix* with a superposition argument, which depends, in turn, upon a pair of lemmas (§3 and §4). Of the three proofs, Gauss' comes closest to Hilbert's standards of rigor, but Lobachevski's is by far the most elegant.

---

[1] This can be done. See, for example, Ch. 9 of Greenberg.

Theory of Parallels 19

In a rectilinear triangle, the sum of the three angles cannot exceed two right angles.

Angle Sum and the Parallel Postulate

In proposition I.32 of the *Elements*, Euclid demonstrates that the angle sum of every triangle is π. It was well-known in Lobachevski's day that this theorem is *logically equivalent* to the parallel postulate. In fact, the equivalence of the two statements was such common knowledge that Lobachevski apparently felt no need to prove it, or even to mention it explicitly, in *The Theory of Parallels*. It is not a hard equivalence to establish.

Claim 1. Given Euclid's first four postulates, the parallel postulate holds if and only if the sum of the angles in every triangle is π.

Proof. $\Rightarrow$) Euclid I.32.

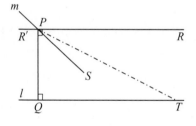

$\Leftarrow$) Assuming that every triangle has an angle sum of π, we shall prove that Playfair's axiom holds.

Let l be a line, and P be a point not on it. From P, drop the perpendicular PQ to l and construct $R'R$ through P perpendicular to PQ. $R'R$ does not intersect l, by Euclid I.28. We shall show that every other line through P does intersect l, and thus establish Playfair's axiom.

To this end, let m be any other line through P. Clearly, m enters either $\angle QPR'$ or $\angle QPR$. We will assume without loss of generality that the latter occurs. Let S be any point on the portion of m lying within $\angle QPR$. Let T be any point on l such that $\angle QTP < \angle SPR$.[1]

By hypothesis, the right triangle $\triangle PQT$ has angle sum π. Thus, $\angle QTP$ is the complement of $\angle TPQ$. Since $\angle TPR$ is also the complement of $\angle TPQ$, it follows that $\angle TPR = \angle QTP < \angle SPR$, which implies that PS enters triangle $\triangle PQT$ through vertex P. What goes in must come out, and PS must intersect QT by the crossbar theorem (see the footnote in TP 17). Hence, m intersects l, as claimed.

This establishes Playfair's axiom, which, in turn, implies the parallel postulate (see the proof of Claim 1 in the notes on TP 16). ■

---

[1] It sounds reasonable that there should be such a point, but this requires proof. Like the parallel postulate, the statement that T exists is an assertion that something will happen at an indefinitely large distance. We can prove that T exists; Lobachevski does this in TP 21. Skeptical readers may turn there immediately for a proof, which employs none of the intervening results.

This raises the question as to what, if anything, we may say about a triangle's angle sum without invoking the parallel postulate. Euclid provides a partial answer. In *Elements* I.17 (a neutral theorem), he demonstrates that any pair of angles in any triangle sums to less than π. From this, we can easily obtain a neutral result on angle sum: the angle sum of any triangle must be strictly less than $(3/2)\pi$.[2]

Lobachevski sharpens the upper bound in TP 19, proving that in neutral geometry, the angle sum can never exceed π. This is the sharpest possible bound on angle sum, since we can actually attain the value π by assuming the parallel postulate along with the neutral axioms.

The proof proceeds by *reductio ad absurdum*. We shall assume the existence of a triangle with angle sum $\pi + \alpha$, where α is some positive real number, and reason to a contradiction.

The Siphon Construction

Suppose that the sum of the three angles in a triangle is $\pi + \alpha$.

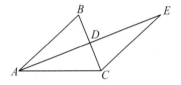

Bisect the smallest side BC at D, draw the line AD, make its extension DE equal to AD, and draw the straight line EC. In the congruent triangles $\triangle ADB$ and $\triangle CDE$ (TP 16 and TP 10), we have $\angle ABD = \angle DCE$ and $\angle BAD = \angle DEC$. From this, it follows that the sum of the three angles in $\triangle ACE$ must also be $\pi + \alpha$. We note additionally that $\angle BAC$, the smallest angle of $\triangle ABC$ (TP 9), has been split into two parts of the new triangle $\triangle ACE$; namely, the angles $\angle EAC$ and $\angle AEC$.

From the original triangle, $\triangle ABC$, Lobachevski produces a second, $\triangle ACE$. The construction acts a kind of siphoning process on the original triangle: it empties the content of its two largest angles (those at B and C) into a single angle of $\triangle ACE$ (the one at C), and drains the content of its smallest angle ($\angle BAC$) into the two remaining angles of $\triangle ACE$ (those at A and E). The resulting triangle has the same angle sum ($\pi + \alpha$) as the original triangle, but comprises one conspicuously large angle and two very small angles. In particular, $\triangle ACE$'s smallest angle must be less than or equal to $(\angle BAC)/2$. (Proof: If not, then the sum of $\triangle ACE$'s two small angles would exceed $\angle BAC$, contradicting the siphon construction, which dictates that their sum will equal $\angle BAC$.)

To recapitulate, the siphon construction produces a new triangle from an old one: in doing so, it preserves the original's angle sum, but reduces its *smallest* angle to less than (or equal to) half of its original size.

---

[2] Proof: In an arbitrary triangle, label the largest angle α, the middle angle β, and the smallest angle γ. By I.17, we know that $\beta < \pi - \alpha$. We also know that $\gamma < \pi/2$ (if not, then $\beta + \gamma = \gamma + \gamma = \pi$, contrary to I.17). Thus, $\alpha + \beta + \gamma < [\alpha + (\pi - \alpha) + \pi/2] = (3/2)\pi$, as claimed.

The Siphon Iterated: The Proof Concluded

Continuing in this manner, always bisecting the side lying opposite the
smallest angle, we eventually obtain
a triangle in which $\pi + \alpha$ is the sum
of the three angles, two of which are
smaller than $\alpha/2$ in absolute magni-

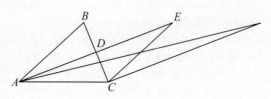

tude. Since the third angle cannot exceed π, α must be either zero or negative.

We may streamline Lobachevski's argument by invoking Euclid I.17 as follows.

We began with $\triangle ABC$, a triangle whose smallest angle is $\angle BAC$. Applying the siphon construction, we produced $\triangle ACE$, whose smallest angle is at most $(\angle BAC)/2$. If we iterate the siphon n times, we will obtain a triangle whose smallest angle is at most $(\angle BAC)/2^n$. We can force this value to be as small as we like by taking n sufficiently large. In particular, we can iterate the procedure until we produce a triangle whose smallest angle is strictly less than α. Since the siphon preserves angle sum, this last triangle will have the same angle sum as the first: $\pi + \alpha$. Consequently, the sum of its two remaining angles must exceed π, contradicting Euclid I.17. Having reached a contradiction, we conclude that a triangle's angle sum cannot exceed π in neutral geometry, as claimed.

The theorem we have just examined is usually called the *Saccheri-Legendre theorem*, for reasons we shall now explain.

Legendre

"It even seems to me that Legendre entered many times
on the same path that I have succeeded in traversing. But
his prejudices in favor of the ideas generally received
until then have, without doubt, always led him to stop at
conclusions which would not be admissible in the new
theory."

— Lobachevski[3]

The wonderful "siphon" proof was actually discovered by Legendre, for whom it represented but one half of a much grander achievement. The siphon proof shows that the angle sum of a triangle cannot exceed π. In several editions of his *Éléments de Géométrie*, Legendre followed this proof with a disturbingly convincing demonstration that the angle sum of a triangle cannot fall short of π either. Concluding that angle sum must therefore equal π, Legendre claimed that he had proved the parallel postulate.

Regarding Legendre's faulty proof, Jeremy Gray has written, "In spotting the flaw you will discover more about the alien nature of non-Euclidean geometry than by following any texts."[4]

---

[3] Lobachevski, *New Principles of Geometry*, pp. 5–6.

[4] Gray, *Ideas of Space*, pg. 81.

I agree, and accordingly reproduce Legendre's proof (in the translation from Laubenbacher & Pengelley) for the reader's edification.

In any triangle, the sum of the three angles is equal to two right angles.

Having already proved that the sum of the three angles of the triangle cannot exceed two right angles, it remains to prove that the same sum cannot be smaller than two right angles.

Let ABC be the proposed triangle, and let, if possible, the sum of its angles $= 2P - Z$, where P denotes a right angle, and Z is whatever quantity by which one supposes the angle sum is less than two right angles.

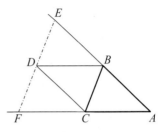

Let A be the smallest of the angles in triangle ABC; on the opposite side BC make the angle $BCD = ABC$, and the angle $CBD = ACB$; the triangles BCD, ABC will be equal, by having an equal side BC adjacent to two corresponding equal angles. Through the point D draw any straight line EF that meets the two extended sides of angle A in E and F.

Because the sum of the angles of each of the triangles ABC, BCD is $2P - Z$, and that of each of the triangles EBD, DCF cannot exceed $2P$, it follows that the sum of the angles of the four triangles ABD, BCD, EBD, DCF does not exceed $4P - 2Z + 4P$, or $8P - 2Z$. If from this sum one subtracts those of the angles at B, C, D, which is $6P$, because the sum of the angles formed at each of the points B, C, D is $2P$, the remainder will equal the sum of the areas of triangle AEF; therefore the sum of the angles of triangle AEF does not exceed $8P - 2Z - 6P$, or $2P - 2Z$. Thus while it is necessary to add Z to this sum of the angles in triangle ABC in order to make two right angles, it is necessary to add at least $2Z$ to the sum of the angles of triangle AEF in order to likewise make two right angles.

By means of the triangle AEF one constructs in like manner a third triangle, such that it will be necessary to add at least $4Z$ to the sum of its three angles in order for the whole to equal two right angles; and by means of the third one constructs similarly a fourth, to which it will be necessary to add at least $8Z$ to the sum of its angles, in order for the whole to equal two right angles, and so forth.

Now, no matter how small Z is in relation to the right angle P, the sequence $Z, 2Z, 4Z, 8Z$, etc., in which the terms increase by a doubling ratio, leads before long to a term equal to $2P$ or greater than $2P$. One will consequently then reach a triangle to which it will be necessary to add to the sum of its angles a quantity equal to or greater than $2P$, in order for the total sum to be just $2P$. This consequence is obviously absurd; therefore the hypothesis with which one started cannot manage to continue to exist, that is, it is impossible that the sum of the angles of triangle ABC is less than two right angles; it cannot be greater by virtue of the preceding proposition; thus it is equal to two right angles.

The flaw occurs early, when Legendre asks his reader to draw a straight line through D meeting the two sides of the angle. This sounds innocent enough, but the statement, "through any point in the interior of an angle, a line may be drawn intersecting both rays of the angle" turns out to be logically equivalent to the parallel postulate, as we shall now demonstrate. Thus, Legendre's argument begs the question, unconsciously assuming that which he claims to prove.

Claim 2. Given Euclid's first four postulates, the parallel postulate holds if and only if "Legendre's assertion" holds. ("Legendre's assertion": through any point in the interior of an angle whose measure is less than π, a line may be drawn intersecting both rays of the angle.)

Proof. $\Leftarrow$) Legendre's flawed proof demonstrates that "Legendre's assertion" implies that the sum of the angles in any triangle equals two right angles, which in turn, implies the parallel postulate (by Claim 1).

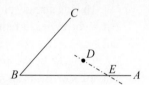

$\Rightarrow$) Suppose the parallel postulate holds. Given an angle $\angle ABC$, we let θ be its measure, and D be an arbitrary point in its interior. On ray BA, choose any point E such that $\angle DEB < \pi - \theta$.[5] Draw line ED. Since lines BC and ED meet BE at angles whose sum is $(\angle ABC + \angle DEB) < (\theta + (\pi - \theta)) = \pi$, the parallel postulate implies that these lines will intersect one another. Thus, ED is a line through D that cuts both rays of angle $\angle ABC$. Hence, "Legendre's assertion" holds. ■

Interestingly, Legendre was not the first to base an alleged proof of the parallel postulate on "Legendre's assertion". In the early 9th century, Abbās ibn Sa'īd al-Jawharī based his own flawed proof of the postulate upon "Legendre's assertion". Unlike Legendre, al-Jawharī provided a proof of this assertion, but this second proof relied upon yet another statement equivalent to the parallel postulate: parallel lines are everywhere equidistant. To his credit, al-Jawharī tried to prove this as well, but it was in this third proof that he made what Rosenfeld characterized as a "crude logical error".[6]

Before moving on to Saccheri, it is worth mentioning that Legendre actually discovered two different proofs that the angle sum cannot exceed π, of which the siphon argument was the second.[7] Finally, one may find the siphon construction, though used to different effect, in Euclid's proof of the exterior angle theorem (*Elements* I.16). Presumably, the close attention that Legendre paid to Euclid while preparing his own *Éléments* gave him the idea for his siphon argument.

Saccheri

> "It is manifest to all geometers that the hypothesis of the right angle alone is true."
>
> —Saccheri[8]

Of all the attempts to prove the parallel postulate, the most heroic—if ultimately Quixotic—was the glorious effort of Gerolamo Saccheri (1667–1733). He was a Jesuit priest, a professor of mathematics and philosophy at the University of Pavia, and most significantly for us, he was the author of *Euclides ab Omni Naevo Vindicatus* (*Euclid Freed of Every Flaw*), which he published in the last year of his life.

---

[5] As mentioned in an earlier footnote, Lobachevski proves in TP 21 that this can be done in neutral geometry.

[6] Rosenfeld, p. 49.

[7] A translation of the first proof, somewhat lengthier than the siphon argument, but clever in its own right, can be found in Laubenbacher & Pengelley, pp. 27–8. A "retelling" of this first proof is in Bonola, pp. 55–56.

[8] Saccheri, p. 61.

In accordance with long-standing custom among postulate provers, Saccheri commences his book with a eulogy for Euclid's *Elements* and a concomitant acknowledgment of the few minor imperfections of that masterpiece:

> No one who has learned mathematics can fail to be aware of the extraordinary merit of Euclid's *Elements*. I call as expert witnesses Archimedes, Apollonius, Theodosius, and the almost innumerable other writers on mathematics up to the present who make use of Euclid's *Elements* as a long established and unshakable foundation. But this great prestige of the *Elements* has not prevented many ancient as well as modern geometers, including many of the most distinguished, from claiming that they have found certain blemishes in this beautiful work, which cannot be too highly praised. Three such blemishes have been cited, which I now give.[9]

Not surprisingly, the first and most significant blemish in Saccheri's list is the existence of the parallel postulate as such; he devotes the entire first part of his book to removing this first blemish, by proving the postulate as a theorem. (The other blemishes concern definitions in the theory of proportions, which need not concern us here.)

Saccheri's purported proof of the postulate was surely the longest ever penned. Covering over 100 pages, it comprises 39 propositions, 5 lemmas, 23 corollaries, and 19 scholia. Its great length was largely a product of Saccheri's unusual strategy. The point of departure for his work is a simple figure, which one constructs as follows: from the endpoints of a base line segment AB, erect perpendiculars AC and BD of the same length, and join their endpoints. The resulting figure is known today as a "Saccheri quadrilateral". Saccheri proves that the figure's summit angles (those at C and D) must equal one another. Furthermore, he proves that Saccheri quadrilaterals throughout the plane exhibit a certain uniformity: if one such quadrilateral has right angles at its summit, then all will; if one has obtuse angles at its summit, then all will; and if one has acute angles at its summit, then all will. He calls these possibilities the *hypothesis of the right angle* (HRA), *hypothesis of the obtuse angle* (HOA), and *hypothesis of the acute angle* (HAA) respectively.

The parallel postulate is equivalent to the HRA. Thus, to prove it, Saccheri sought to demonstrate that the HOA and HAA both lead to logical absurdities. Initially, his plan appears feasible; he successfully explodes the HOA in his 14th proposition and celebrates this first victory in zesty Latin[10] before commencing "a lengthy battle against the hypothesis of the acute angle, which alone opposes the truth. . . "[11] He thus assumes the HAA, fully intending to drive it toward its own destruction as well. This attempt led him through a lengthy sequence of deductions; his increasingly strange results seemed to contradict experience without actually contradicting logic. Recognizing that this was insufficient for his purposes, Saccheri had no choice but to plunge still deeper into the world of the HAA in quest of logical absurdity.

Saccheri's desperate quest was doomed to fail: with hindsight, we know that the HAA does *not* lead to a contradiction. The strange propositions that Saccheri established under his HAA fever-dream were not, as he had fancied, mere hallucinatory stepping-stones that would lead him

---

[9] Saccheri, p. 245.

[10] "*Hypothesis anguli obtusi est absolute falsa, quia se ipsam destruit.*" ("The hypothesis of the obtuse angle is absolutely false, because it destroys itself.") Saccheri, p. 61.

[11] *ibid.* p. 13.

to a logical contradiction, but rather, honest theorems of imaginary geometry, which Lobachevski would rediscover a century later. Saccheri beheld a new world, but failed to recognize it.

Indeed, he finally deluded himself into claiming victory over the HAA. The ostensible contradiction that prompted this was so patently bogus that Saccheri, whose logical acumen was otherwise profound, must have accepted it out of sheer mental exhaustion, if not intellectual terror. One cannot help wishing that he had lived a little longer to reconsider his overhasty conclusion. Though he lost his battle with the HAA, he became, in waging it, the first person in history to make sustained, if unwitting, contact with non-Euclidean geometry.

The Saccheri-Legendre theorem was called *Legendre's first theorem* until the early 20th century.[12] Although Saccheri's *Euclides ab Omni Naevo Vindicatus* seems to have captured the interest of mathematicians at the time of its publication in 1733,[13] it had long since sunk into oblivion by Legendre's day. It was rediscovered in 1889 by Eugenio Beltrami, who introduced it to a generation of mathematicians able to appreciate Saccheri's unsuspecting incursion into the world of non-Euclidean geometry.[14] Within the pages of Saccheri's treatise, these late 19th-century mathematicians discovered a proof of "Legendre's first theorem" that antedated Legendre's own by the better part of a century.

Saccheri's proof is as follows. In his 15th proposition, he demonstrates that HAA, HRA, and HOA, lead respectively to systems in which a triangle's angle sum is always less than π, always equal to π, or always greater than π. Combining this with his destruction of the HOA, Saccheri establishes the theorem that now bears his (hyphenated) name.

Euclid's Second Postulate

Euclid's second postulate (*"To produce a finite straight line continuously in a straight line"* in Heath's translation) is generally taken as an assertion that the plane is *unbounded*; line segments can be extended indefinitely. The proof of the Saccheri-Legendre theorem depends crucially upon this fact, since, in carrying out the siphon construction, we require the ability to extend a segment to twice its length. In spherical geometry, where Euclid's second postulate does not hold (great circles, the "lines" of a sphere, have finite length, thus limiting the amount by which a segment can be extended), Lobachevski's proof of the Saccheri-Legendre theorem is no longer valid. This should not be surprising, since the sum of any spherical triangle's angles is, in fact, always *greater* than π.

---

[12] e.g. Hilbert p. 35. Writing in 1906, Bonola (p. 56) notes, "This theorem is usually, but mistakenly, called *Legendre's First Theorem.*"

[13] Coolidge, p. 70. Coolidge bases this claim upon a 1903 article (written in Italian) by Corrado Segre on the influence of Saccheri upon subsequent writers concerned with the parallel postulate.

[14] There is some irony in the fact that Beltrami, who revived the long-lost work and name of Saccheri, also demonstrated that imaginary geometry (the HAA) is as consistent as Euclidean geometry (the HRA), thus proving definitively that Saccheri's quest was hopeless.

Theory of Parallels 20

If the sum of the three angles in one rectilinear triangle is equal to two right angles, the same is true for every other triangle.

All for One and One for All

This proposition, like the previous one, played an important role in Legendre's purported proofs of the parallel postulate. It is sometimes called *Legendre's second theorem*, but a pleasantly literary alternative, *The Three Musketeers Theorem*,[1] has gained favor in recent years. Whatever its name, Legendre was not the first to prove it. Saccheri preceded him once again. Nonetheless, Legendre's proof, which Lobachevski largely follows, is particularly elegant. The enormous popularity of Legendre's *Éléments* was due, in no small measure, to his artful proofs, which seem even to invite names: we have seen his "siphon construction" in TP 19, and we will now examine his "domino proof" of the Three Musketeers Theorem, as retold by Lobachevski.

If a Triangle with Angle Sum π Exists, Then a Right Triangle with Angle Sum π Exists.

If we suppose that the sum of the three angles in triangle $\triangle ABC$ is equal to π, then at least two of its angles, A and C, must be acute. From the third vertex, B, drop a perpendicular p to the opposite side, AC. This will split the triangle $\triangle ABC$ into two right triangles. In each of these, the angle sum will also be π: neither angle sum can exceed π (TP 19), and the fact that the right triangles comprise triangle $\triangle ABC$ ensures that neither angle sum is less than π.

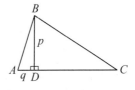

---

[1] Interestingly, Alexandre Dumas *père*, author of *The Three Musketeers*, has a curious connection to the history of mathematics. He was present at the banquet of French revolutionaries at which the great algebraist Evariste Galois apparently declared his intention, while standing on a table with a drawn dagger, to kill the king. Moreover, in his memoirs, Dumas names Pescheux d'Herbinville as the man who killed the twenty-year-old Galois in a duel. Dumas' writings provide the only evidence pointing to d'Herbinville; whether Dumas correctly identified Galois' killer is still a matter of debate. See the article by Tony Rothman listed in the bibliography.

Since angles A and C are acute, the foot of the perpendicular will land in the interior of segment AC^2. Hence, the perpendicular splits $\triangle ABC$ into a pair of right triangles. We shall verify that each right triangle has angle sum π by proving a little lemma, which gives a slightly more general result.

Claim 1. Given a triangle with angle sum π, if we join one of its vertices to a point on the opposite side, both of the resulting subtriangles have angle sum π as well.

Proof. Suppose that *angle sum*$(\triangle ABC) = \pi$. Join B to a point D on the opposite side to produce two subtriangles, $\triangle ADB$ and $\triangle CDB$. If we add up all six angles of the subtriangles, we clearly obtain all three angles of the original triangle, plus a pair of supplementary angles at D. That is,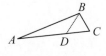

$$angle\ sum(\triangle ADB) + angle\ sum(\triangle CDB) = angle\ sum(\triangle ABC) + \pi = 2\pi.$$

By TP 19, neither term on the left hand side can exceed π. Hence, the equality can hold only if both terms equal π. That is, both subtriangles have angle sum π, as claimed. ∎

Thus, the existence of one arbitrary triangle with angle sum π implies the existence of a *right* triangle with angle sum π. The chain of dominoes has begun to fall.

If a Right Triangle with Angle Sum π Exists, Then Arbitrarily Large *Right Triangles of Angle Sum π Exist.*

In this way, we obtain a right triangle whose legs are p and q; from this we can obtain a quadrilateral whose opposite sides are equal, and whose adjacent sides are perpendicular (Fig. 6). By repeated application of this quadrilateral, we can construct another with sides np and q, and eventually a quadrilateral $EFGH$, whose adjacent sides are perpendicular, and in which $EF = np, EH = mq, HG = np,$ and 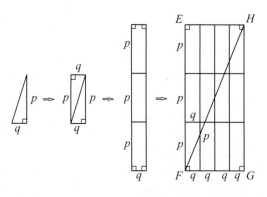 $FG = mq$, where m and n can be any whole numbers. The diagonal FH of such a quadrilateral divides it into two congruent right triangles, $\triangle FEH$ and $\triangle FGH$, each of which has angle sum π.

From a right triangle whose angle sum is π, Lobachevski (following Legendre) produces a rectangular brick. Using copies of the brick, he builds arbitrarily large rectangular walls. Drawing a diagonal of such a wall yields an arbitrarily large right triangle with angle sum π. We now explain why this works.

---

[2] *Proof.* If the foot of the perpendicular fell outside the segment, say closer to A than to C, then $\triangle BDA$ would have a right angle at D and an obtuse angle at A, contradicting TP 19, the Saccheri-Legendre Theorem.

Since triangle $\triangle CDB$ has angle sum π, its two acute angles are complementary. Joining two copies at their hypotenuses therefore yields our first $p \times q$ rectangular brick, a "quadrilateral whose opposite sides are equal, and whose adjacent sides are perpendicular."[3]

Stacking such bricks n high and m across, we obtain an $np \times mq$ rectangular wall. With a diagonal, we split it into a pair of right triangles. If we add the angle sums of these triangles together, it is clear that we will obtain 2π, since this is the angle sum of the rectangle they comprise.

Hence, if either triangle's angle sum falls short of π, the other's must exceed π to make up the difference, which would contradict the Saccheri-Legendre Theorem (TP 19). Thus neither triangle's angle sum is less than π. Since neither angle sum can be greater than π either (TP 19 again), both angle sums must equal π.

Thus, given just one triangle (possibly a tiny one) with angle sum π, we can construct arbitrarily large right triangles with angle sum π. Specifically, we can construct them to have legs of lengths np and mq, for any whole numbers m and n whatsoever. Given any specified length, we can therefore construct a right triangle of angle sum π, whose legs exceed that length.

If Arbitrarily Large Right Triangles of Angle Sum π Exist, Then Every Right Triangle has Angle Sum π.

The numbers m and n can always be chosen so large that any given right triangle $\triangle JKL$ can be enclosed within a right triangle $\triangle JMN$, whose legs are $NJ = np$ and $MJ = mq$, when one brings their right angles into coincidence. Drawing the line LM yields a sequence of right triangles in which each successive pair shares a common side.

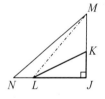

The triangle $\triangle JMN$ arises as the union of the triangles $\triangle NML$ and $\triangle JML$. The angle sum exceeds π in neither of these; it must, therefore, equal π in each case in order to make the composite triangle's angle sum equal to π. Similarly, the triangle $\triangle JML$ consists of the two triangles $\triangle KLM$ and $\triangle JKL$, from which it follows that the angle sum of $\triangle JKL$ must equal π.

Given any right triangle $\triangle JKL$, we can construct, by the technique described in the previous section, a right triangle $\triangle JMN$ whose angle sum is π, and whose legs are longer than those of $\triangle JKL$. When their right angles coincide, it is obvious that $\triangle JMN$ will enclose the given triangle $\triangle JKL$. (One can give a Hilbert-style proof of this fact, but I will omit this here.) Lobachevski alludes to a "sequence of right triangles": $\triangle JMN$, $\triangle JML$, $\triangle JKL$. Applying Claim 1 to each successive pair of triangles in the sequence brings us to the desired conclusion.

Since $\triangle JMN$ has angle sum π, its subtriangle $\triangle JML$ has angle sum π as well.

Since $\triangle JML$ has angle sum π, its subtriangle $\triangle JKL$ does also.

---

[3] Strictly speaking, "joining triangles" is not a well-defined operation. To be more precise, we can construct a rectangle $BDCE$ from triangle $\triangle BDC$ by choosing a point E such that $\angle CBE = \angle DCB$ (Euclid I.23) and $BE = DC$ (Euclid I.2). By the SAS criterion (Euclid I.4), we have $\triangle BDC \cong \triangle CEB$. Hence, $EC = BD$. Since $BE = DC$ by construction, the opposite sides of quadrilateral $BDCE$ are equal. Moreover, $\angle DBE = \angle DBC + \angle CBE = \angle DBC + \angle BCD = \pi/2$. Similarly, $\angle DCE = \pi/2$. Since the remaining two angles of $BDCE$ are obviously right angles, we have also confirmed that all pairs of adjacent sides in the quadrilateral are perpendicular.

Thus far, we have shown that the existence of just one triangle with angle sum π implies that every *right* triangle has angle sum π. Only one more domino remains to fall.

If Every Right Triangle has Angle Sum π, Then Every Triangle has Angle Sum π.

In general, this must be true of every triangle since each triangle can be cut into two right triangles. Consequently, only two hypotheses are admissible: the sum of the three angles either equals π for all rectilinear triangles, or is less than π for all rectilinear triangles.

Let T be an arbitrary triangle. We may dissect it into two right triangles, T_1 and T_2, by dropping a perpendicular from the appropriate vertex. From the figure, we see that

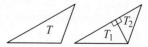

$$angle\ sum(T) = angle\ sum(T_1) + angle\ sum(T_2) - \pi.$$

If all right triangles have angle sum π, it follows that

$$angle\ sum(T) = \pi + \pi - \pi = \pi.$$

Thus, T has angle sum π as claimed. This completes the domino proof of the Three Musketeers Theorem. To recapitulate, when we assume that one triangle has angle sum π, a chain of deductions leads us inexorably to the conclusion that *all* triangles have angle sum π. This has an important consequence for imaginary geometry.

Claim 2. In imaginary geometry, the angle sum of every triangle is strictly less than π.

Proof. Since the parallel postulate is false in imaginary geometry, it harbors at least one triangle whose angle sum is *not* π. (TP 19 notes, Claim 1). Thus, by the Three Musketeers Theorem, *no* triangles can have angle sum π in imaginary geometry. Thus, the angle sum of every triangle in imaginary geometry is strictly less than π (by the Saccheri-Legendre Theorem). ∎

We have reached the end of Lobachevski's TP 20, but much remains to be discussed. Our demonstration that every triangle in imaginary geometry has angle sum less than π immediately spawns questions. Foremost among them: do all imaginary triangles possess the same angle sum? If so, what is it? If not, can we discover a law that describes their variation? Johann Heinrich Lambert was the first to answer these questions, and it is to his work that we turn next.

Lambert

"Lambert may be compared in a sense with Moses, for he saw more of the promised land of the new geometry than anyone before him, and knew that he had not proved it self-contradictory, but . . ."

—Jeremy Gray[4]

---

[4] Fauvel & Gray, p. 509.

The far-reaching intellectual achievements of Johann Heinrich Lambert (1728–1777) span the disciplines of mathematics, physics, astronomy, and philosophy. His was among the greatest of 18th-century scientific minds. Immanuel Kant drafted a dedication of his own *Critique of Pure Reason* to Lambert, but by the time Kant's masterpiece was ready to publish, Lambert had died. Some of his better known mathematical accomplishments include proofs that π and e are irrational numbers, the introduction of the hyperbolic functions, and significant work in mathematical cartography.

The vexed subject of the parallel postulate attracted Lambert's attention. He left one important work on this topic, his *Theorie der Parallel-Linien* (written in 1766, but published posthumously in 1786, by Johann Bernoulli III). Although Lambert never explicitly mentions Saccheri, he was probably familiar with his work; he does explicitly refer to a 1763 dissertation of G.S. Klügel, which summarized and criticized various proofs of the parallel postulate, including Saccheri's *Euclid Freed from Every Flaw*.

Lambert's own contribution to the "pre-history" of non-Euclidean geometry proceeds along lines very similar to those that Saccheri had followed 33 years earlier. He considers a quadrilateral with three right angles (now often called a "Lambert quadrilateral"), and establishes the following neutral result about its fourth angle: if this fourth angle of any particular Lambert quadrilateral is right, obtuse, or acute, then the fourth angle of *every* Lambert quadrilateral will be right, obtuse, or acute, respectively. These three possibilities, which Lambert calls the *first, second, and third hypotheses*, are entirely equivalent to Saccheri's HRA, HOA, and HAA. Like Saccheri, Lambert sets out to destroy the second and third hypotheses. The second falls easily, but the third (the HAA) remains stubbornly resistant, prompting a long excursion into a strange geometric landscape. Lambert followed the path into imaginary geometry even farther than Saccheri did, thus anticipating several theorems of Lobachevski and Bolyai.

In the pages that follow, we shall find occasion to mention several of Lambert's results. Of interest here is the fact that different triangles in imaginary geometry can have different angle sums. Saccheri never explicitly mentions this fact, but in section 81 of Lambert's *Theorie der Parallellinien*, one finds a demonstration that under the "third hypothesis" (equivalently, in imaginary geometry) if one triangle is inscribed in another, then the larger triangle will have a smaller angle sum. A synopsis of his proof follows[5].

Claim 3. In imaginary geometry, if $\triangle ABC$ is inscribed in $\triangle XYZ$, then angle sum($\triangle XYZ$) < angle sum($\triangle ABC$).

Proof. The inscribed triangle splits $\triangle XYZ$ into four sub-triangles: $\triangle ABC$, $\triangle ACY$, $\triangle BXC$, and $\triangle ZBA$. Let the angle sums of these triangles be $\pi-\alpha$, $\pi-\beta$, $\pi-\gamma$, and $\pi-\delta$ respectively. The figure shows that $\triangle XYZ$'s angle sum can be found by adding up the four angle sums of the sub-triangles and subtracting 3π. (The 3π accounts for the three straight angles at A, B, and C, which are composed of angles of the

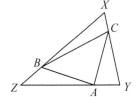

―――――――
5 For an English translation of Lambert's work on area and angle sum, see Fauvel & Gray, pp. 518–20. The complete German text of Lambert's *Theorie der Parallellinien* is in Engel & Stäckel, pg. 152–207. As of this writing, this text can be accessed online through the World Digital Mathematics Library (http://www.wdml.org/)

sub-triangles that do not figure into $\triangle XYZ$'s angle sum.) That is,

$$angle\ sum(\triangle XYZ) = [(\pi - \alpha) + (\pi - \beta) + (\pi - \gamma) + (\pi - \delta) - 3\pi] = \pi - (\alpha + \beta + \gamma + \delta),$$

which is less than $\pi - \alpha$, the angle sum of the inscribed triangle. (It is also less then the angle sum of any of the other three sub-triangles, for that matter.) ■

Lambert generalized this result, arguing that for any two triangles, the one with greater area will always have a smaller angle sum. In his sketchy justification of this claim, Lambert suggests that variations on the proof of Claim 3 will allow us to demonstrate the truth of the inequality for any given pair of triangles.[6] He then made this result more precise, arguing that a triangle's *angle defect* (the amount by which its angle sum falls short of π) is in fact directly proportional to its area. Gauss crafted a beautiful proof of this theorem; I shall describe it in the notes to TP 33. For now, I shall content myself with one small step in this direction. Like area, angle defect turns out to be an *additive* quantity, in the sense indicated in the following claim.

Claim 4. If we split a triangle into two subtriangles by joining a vertex to a point on the opposite side, then the angle defects of the subtriangles add up to the angle defect of the original triangle.

Proof. Let $\triangle ABC$ be the original triangle, with subtriangles $\triangle ABD$ and $\triangle ACD$, as in the figure. If we label angles as in the figure, then the defects of the three triangles are:

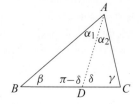

$$Angle\ defect(\triangle ABD) = \pi - [\alpha_1 + \beta + (\pi - \delta)].$$
$$Angle\ defect(\triangle ADC) = \pi - [\alpha_2 + \delta + \gamma].$$
$$Angle\ defect(\triangle ABC) = \pi - [\alpha_1 + \alpha_2 + \beta + \gamma].$$

A simple calculation shows that the sum of the first two defects on this list is equal to the third, as claimed. ■

This additivity implies that in imaginary geometry (where angle defect is always positive), the original triangle's defect will exceed either subtriangle's defect. An easy corollary, which we will use shortly, follows.

Corollary. *In imaginary geometry, if a chord drawn in a triangle splits the triangle into a subtriangle and a quadrilateral, then the original triangle's angle defect is greater than the subtriangle's defect.*

Proof. Let $\triangle ABC$ be the original triangle, DE the chord, and $\triangle ADE$ the subtriangle. Draw line DC. Appealing to Claim 4 twice, we obtain

$$Angle\ defect(\triangle ABC) > angle\ defect(\triangle ADC)$$
$$> angle\ defect(\triangle ADE).$$

Thus, the original triangle's defect exceeds that of its subtriangle, as claimed. ■

------

[6] "I shall not prove this theorem completely here," he writes, "... rather I shall give only so much of the proof as will enable the rest of it to be understood overall." Lambert's claim is true, but some subtle complications do arise when one tries to vary the proof of Claim 3 to handle cases in which the triangle of smaller area does not fit inside the triangle of larger area.

While pursuing his "third hypothesis", Lambert obtained many further bewildering results, but unlike Saccheri, he never succumbed to the illusion that he had found a logical contradiction amongst them. Presumably dissatisfied with the inconclusive nature of his studies, he never published his "incomplete proof of the parallel postulate" in his lifetime. In his cool, dispassionate view of the third hypothesis, he stands in marked contrast to his great predecessor Saccheri, who attacked the HAA as if it were his personal enemy.

Similarity and the Parallel Postulate: Wallis

> "In time, those Unconscionable Maps no longer satisfied, and the Cartographers Guilds struck a map of the Empire whose size was that of the Empire, and which coincided point for point with it . . . "
> —Jorge Luis Borges, "On Exactitude in Science"

Lambert's results have a remarkable consequence: since increasing the size of a triangle decreases its angle sum, it follows that in imaginary geometry, *similar, non-congruent triangles cannot exist.* Dilating a figure invariably distorts its angles, so in a world governed by imaginary geometry, photography would be an inherently surrealist art, as Marvin Greenberg has aptly noted.[7] The familiar AAA-similarity criterion for Euclidean triangles (Euclid VI.4) disappears; in imaginary geometry, AAA is a *congruence* criterion.

Claim 5. In imaginary geometry, AAA is a congruence criterion.

Proof. Let $\triangle ABC$ and $\triangle A'B'C'$ be triangles whose corresponding angles are equal.

Suppose, by way of contradiction, that the triangles are *not* congruent.

Then none of their corresponding sides are equal; for if they had a pair of equal corresponding sides, the triangles would be congruent by the ASA-criterion, contrary to hypothesis. Consequently, one triangle contains (at least) two sides longer than those that correspond to them on the other triangle. Suppose then that $AB > A'B'$ and $AC > A'C'$. Let B'' and C'' be points on AB and AC respectively such that $AB'' = A'B'$ and $AC'' = A'C'$.

Since $\triangle A'B'C' \cong \triangle AB''C''$ (SAS), we have $\angle B' = \angle B''$ and $\angle C' = \angle C''$.

The corollary to Claim 4 tells us that $(\angle B + \angle C) < (\angle B'' + \angle C'')$.

Combining these last two facts yields $(\angle B + \angle C) < (\angle B' + \angle C')$.

On the other hand, all corresponding angles of $\triangle ABC$ and $\triangle A'B'C'$ are equal, so $(\angle B + \angle C) = (\angle B' + \angle C')$.

We have reached a contradiction. Hence, $\triangle ABC \cong \triangle A'B'C'$, as claimed. ∎

In a lecture given July 11, 1663 at Oxford University, John Wallis demonstrated that the existence of similar, non-congruent figures is logically equivalent to the parallel postulate. Wallis' critical examination of the postulate appears to have been the first by a European since ancient times. He was inspired by an Arabic work on parallels, which had been published in Rome in 1594, and attributed to the 13th century mathematician, Nasir Eddin al-Tusi. It has since been

---

[7] Greenberg, p. 151

demonstrated that it was actually written after al-Tusi's death in 1274. Rosenfeld considers it "very likely" that it was written by al-Tusi's son, Sadr al-Din.

Wallis stated that it would be reasonable to assume that "to every figure there is always a similar one of arbitrary size"; since Euclid's 3rd postulate asserts the existence of circles of arbitrary size, "it is as practicable to make this assumption for an arbitrary figure as for circles."[8] Although an appealing justification, this is somewhat misleading, since the existence of arbitrarily large circles does not imply that *similar* circles of arbitrary size exist. In fact, the omni-similarity of Euclidean circles (which ensures that the ratio of circumference to diameter is constant[9], thus justifying the usual definition of π) is not a consequence of Euclid's third postulate, but of his parallel postulate. Wallis, however, avoids circular arguments. Unlike many of his successors (such as Legendre), he never claimed to have derived the parallel postulate from Euclid's first four; he merely showed that it was equivalent to his own proposed postulate, which, he argued, was a somewhat less offensive alternative. A sketch of Wallis' equivalence proof follows.

Claim 6. *Given Euclid's first four postulates, the parallel postulate holds if and only if Wallis' Postulate (to every figure there is always a similar one of arbitrary size) holds.*

Proof. $\Rightarrow$) Euclid VI.25

$\Leftarrow$) Suppose that Wallis' postulate holds. Let m and n be lines intersected by a third line, l, in such a way that the sum of the interior angles that m and n make on one side of l is less than π.

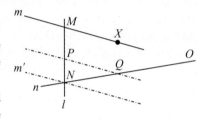

Let M and N be the intersections of l with m and n respectively. Let m' be the line through N making the same angle with l as m does, and let X be a point on m, as in the figure. Slide m along l towards N, maintaining its inclination toward l, until it coincides with m'. By the time it completes this journey, the point X will have passed to the other side of line n. Consequently, X must have crossed n at some point Q during its passage. Of course, when X and Q were coincident, the sliding line was also intersecting l at some point. Call this point P. By Wallis' postulate, there is a point O such that $\triangle MNO \sim \triangle PNQ$. Since O must clearly lie on both of the lines m and n, these lines do intersect. Thus, the parallel postulate holds, as claimed. ∎

In fact, it can be shown that Wallis' postulate (and hence the parallel postulate) holds if and only if there exists a single pair of similar, non-congruent triangles. In 1824, Pierre Simon de Laplace, author of the five-volume *Mécanique Céleste*, reiterated Wallis' idea, that a postulate asserting the existence of similar figures is more natural than Euclid's postulate.

> ...the notion of space includes a special property, self-evident, without which the properties of parallels cannot be rigorously established. The idea of a bounded region, e.g., the circle, contains nothing which depends on its absolute magnitude. But if we imagine its radius to diminish, we are brought without fail to the diminution in the same ratio of its circumference and sides of all the inscribed figures. This proportionality

---

[8] Fauvel and Gray, pp. 510–11

[9] For a proof of this oft-stated, but rarely demonstrated theorem, see Moise, pp. 265–268.

appears to me a more natural postulate than that of Euclid, and it is worthy of note that it is discovered afresh in the results of the theory of universal gravitation.[10]

The reservations that most mathematicians felt regarding the parallel postulate were not logical, but aesthetic; their objections had little to do with the truth of the parallel postulate — they *knew* it was true — but much to do with its position in the structure of the *Elements*. It ought to be a theorem, they insisted, not an axiom. While the desired solution to this aesthetic dilemma was to deduce it from the first four postulates, a second-rate alternative was to replace Euclid's fifth postulate, as Wallis or Laplace suggested, with a less objectionable equivalent. Of all the proposed "self-evident" alternatives, the most disarmingly simple must be the one that Alexis Claude Clairaut (1713 – 1765) adopted in his *Éléments de Géométrie*: there exists a rectangle.

Clairaut's Postulate

Claim 7. *Given Euclid's first four postulates, the parallel postulate holds if and only if a rectangle exists.*

Proof. ⇒) Euclid I.46 (where Euclid constructs a square, a special case of a rectangle.)

⇐) If a rectangle exists, let a diagonal split it into two triangles. The angle sums of these triangles add up to 2π (the angle sum of the rectangle). Since neither angle sum can exceed π (by TP 19), each triangle's angle sum must be exactly π. By the Three Musketeers theorem (TP 20), *all* triangles must therefore have angle sums equal to π, a statement we have seen is equivalent to the parallel postulate. ∎

It is interesting to speculate how the history of non-Euclidean geometry might have unfolded had Euclid assumed the existence of a rectangle rather than his parallel postulate. Would succeeding generations have found Clairaut's postulate a "blot on geometry" (as Henry Saville referred to the parallel postulate in 1621)? Could the mere assumption that a rectangle exists have provoked Saccheri's struggle to "free Euclid from every flaw"?

Yes, it could have, and probably would have. Although Clairaut's postulate seems to have much in common with Euclid's third postulate (asserting the existence of circles), closer inspection reveals a fundamental difference. All of Euclid's geometry is built from line segments and circles; the third postulate merely asserts that we have access to the latter. Certainly, as geometric aesthetes, we should never tolerate the naïve assumption that other figures exist; we should prove their existence rigorously by constructing them from preexisting material. In particular, if a rectangle exists, we should be able to construct it from four line segments. Euclid constructs every other figure he uses, so why should a rectangle be any different? Such thoughts would surely have bothered geometers had Euclid based his theory of parallels upon Clairaut's postulate. Similar thoughts would have bred discontent with *any* equivalent form of the parallel postulate.

---

[10] Quoted in Bonola, p. 54

Theory of Parallels 21

From a given point, one can always draw a straight line that meets a given line at an arbitrarily small angle.

The salient feature of this little result is its neutrality; it is valid in both Euclidean and imaginary geometry. Twice, in the notes following TP 19, I committed the venial sin of using this result without having proved it first, referring the reader to the present proposition for its demonstration. I am not, however, guilty of the mortal sin of circular argument; Lobachevski's proof of TP 21 does not require the intervening proposition, TP 20, and thus it could have been given directly after TP 19, prior to the proofs in which I used it.

From the given point A, drop the perpendicular AB to the given line BC; choose an arbitrary point D on BC; draw the line AD; make $DE = AD$, and draw AE. If we let $\alpha = \angle ADB$ in the right triangle $\triangle ABD$, then the angle $\angle AED$ in the isosceles triangle $\triangle ADE$ must be less than or equal to $\alpha/2$ (TP 8 & 19)[1]. Continuing in this manner, one eventually obtains an angle $\angle AEB$ that is smaller than any given angle.

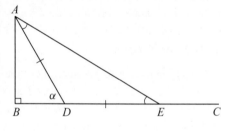

TP 8 (or Euclid I.5, the *pons asinorum*) implies that the base angles $\angle AED$ and $\angle DAE$ of the isosceles triangle $\triangle ADE$ are equal; let β be their common measure. Since this triangle's remaining angle is $\pi - \alpha$, TP 19 (the Saccheri-Legendre theorem) gives $(\pi - \alpha) + \beta + \beta \leq \pi$. Hence, $\beta \leq \alpha/2$, as claimed.

By repeating this construction, allowing E to play the role of D, we find a point F such that $\angle AFD \leq \alpha/4$. With each subsequent iteration, we produce an angle whose measure is less than half the previous one. Repeating this sufficiently many times, we can clearly construct an angle smaller than any specified positive angle given in advance. The point E that occurs in Lobachevski's final sentence is, of course, generally distinct from the E defined earlier in

---

[1] I have corrected an apparent misprint occurring in Lobachevski's text and perpetuated in Halsted's 1891 translation of *TP*. In these sources, Lobachevski cites TP 20 at this point, rather than TP 19. This makes little sense; TP 20 relates the angle sum of one triangle to the angle sums of all triangles—an issue having scarcely anything to do with the present proposition's modest concerns.

the proposition. In geometric writings of this period, the same symbol was sometimes used to represent distinct points that played a similar role in an argument.

Interestingly, Bolyai also makes use of this proposition, which looks so much like an *ad hoc* lemma. He notes the result at the conclusion of §1 in his *Appendix*, directly after defining parallelism.

Theory of Parallels 22

If two perpendiculars to the same straight line are parallel to one another, then the sum of the three angles in all rectilinear triangles is π.

Common Perpendiculars

A theorem of neutral geometry (I.28) guarantees that two lines with a common perpendicular (i.e. two lines perpendicular to a third) will never meet one another. Thus, in Euclidean geometry, lines with a common perpendicular are parallel to one another.

In imaginary geometry, however, non-intersection does not suffice to establish parallelism (see TP 16). In fact, the present proposition demonstrates that in imaginary geometry, lines with a common perpendicular are *not* parallel to one another. (Were they parallel, all triangles would have angle sum π, contradicting the result proved in Claim 2 of the notes for TP 20.)

Accordingly, when we invoke TP 22 in the future, we shall do so in the following equivalent form: ***in imaginary geometry, parallel lines cannot have a common perpendicular***. (That is, given two parallels, there cannot be a third line perpendicular to each of them.)

Proof on the Rack

Let the lines AB and CD (Fig. 9) be parallel to one another and perpendicular to AC. From A, draw lines AE and AF to points E and F chosen anywhere on the line CD such that $FC > EC$. If the sum of the three angles equals $\pi - \alpha$ in the right triangle $\triangle ACE$ and $\pi - \beta$ in triangle $\triangle AEF$, then it must equal $\pi - \alpha - \beta$ in triangle $\triangle ACF$, where α and β cannot be negative. Further, if we let $a = \angle BAF$ and $b = \angle AFC$, then $\alpha + \beta = a - b$.

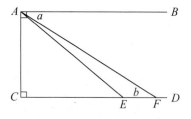

Lobachevski assumes that two parallels AB and CD have a common perpendicular, AC. Proceeding from this assumption, he will produce a triangle with angle sum π; this will imply that *all* triangles have angle sum π (TP 20), which, in turn, will imply that the parallel postulate holds (TP 19 notes, Claim 1). Having accomplished this, he will have shown that parallels can have a common perpendicular only in Euclidean geometry.

His strategy for finding a triangle with angle sum π is worthy of the Spanish Inquisition. He examines two closely related triangles, $\triangle ACE$ and $\triangle AEF$, and while the first can do nothing but watch, he contorts the shape of the second until the first breaks down and confesses that its own angle sum is π.

Lobachevski denotes their angle sums: $\pi - \alpha$ and $\pi - \beta$. He will ultimately show that $\alpha = 0$.

To establish an exploitable relationship between the angle sums of the two triangles, Lobachevski turns to $\triangle ACF$, the larger triangle that they comprise. He computes its angle sum in two different ways.

First, he computes it indirectly, in terms of the angle sums of its constituent sub-triangles. If we denote the respective angle sums of $\triangle ACE$ and $\triangle AEF$ by $\pi - \alpha$ and $\pi - \beta$, then a glance at the figure indicates a method for finding $\triangle ACF$'s angle sum: add the angle sums of its two sub-triangles and subtract π to offset their contributions at E. Thus, by our first computation, $\triangle ACF$'s angle sum is $(\pi - \alpha) + (\pi - \beta) - \pi = (\pi - \alpha - \beta)$.

Second, he computes it directly, in terms of its own angles. These have measure b, $\pi/2$, and $(\pi/2) - a$, so by our second computation, $\triangle ACF$'s angle sum is $(\pi + b - a)$.

Equating the two expressions for $\triangle ACF$'s angle sum, we find that $(\alpha + \beta) = (a - b)$, as claimed.

Having established this relationship, Lobachevski is ready to put $\triangle AEF$ on the rack.

End of the Proof: Confession

By rotating the line AF away from the perpendicular AC, one can make the angle a between AF and the parallel AB as small as one wishes; one reduces the angle b by the same means. It follows that the magnitudes of the angles α and β can be none other than $\alpha = 0$ and $\beta = 0$.

Here, Lobachevski lets F slide down ray CD towards infinity. As this point moves, $\triangle ACE$ remains unaffected, but ray AF rotates about A, causing $\triangle ACE$ (and hence $\triangle ACF$) to be stretched. We shall now consider some limiting behavior of different parts of the figure as F slides to infinity.

Naturally, the limiting position towards which ray AF rotates is the first ray through A that does *not* cut CD: that is, AF approaches the unique parallel to CD passing through A. By hypothesis, this parallel is AB; hence, the angle between AF and AB approaches 0. That is, $a \to 0$ as F goes to infinity.

Moreover, TP 21 immediately tells us that $b \to 0$ as F goes to infinity.

Consequently, it follows that $(a - b) \to 0$ as F goes to infinity.

This is enough: $\triangle ACE$ will now confess.

Since $\alpha \leq (\alpha + \beta) = (a - b)$, which vanishes as F goes to infinity, it follows that α can be made smaller than any positive quantity. Thus, since it cannot be negative (by TP 19), α must be zero. That is, $\triangle ACE$'s angle sum is $(\pi - \alpha) = (\pi - 0) = \pi$, as claimed.

This being the case, *all* triangles have angle sum π, by the Three Musketeers Theorem (TP 20). The proof is over, but Lobachevski has a few more words to add about his unfolding work as a whole.

Recapitulation and Proclamation

From what we have seen thus far, it follows either that the sum of the three angles in all rectilinear triangles is π, while the angle of parallelism $\Pi(p) = \pi/2$ for all lines p, or that the angle sum is less than π for all triangles, while $\Pi(p) < \pi/2$ for all lines p. The first hypothesis serves as the foundation of the ordinary geometry and plane trigonometry.

The second hypothesis can also be admitted without leading to a single contradiction, establishing a new geometric science, which I have named Imaginary Geometry, which I intend to expound here as far as the derivation of the equations relating the sides and angles of rectilinear and spherical triangles.

We have now seen ten statements equivalent to the parallel postulate. Gathering them together, we state a theorem that we have already proved in piecemeal fashion.

Theorem. *Given Euclid's first four postulates, the following are equivalent:*

> *1. Euclid's parallel postulate.*
> *2. Playfair's axiom.* (See TP 16 Notes, Claim 1)
> *3. All triangles have angle sum π.* (See TP 19 Notes, Claim 1)
> *4. One triangle has angle sum π.* (See TP 20)
> *5. The angle of parallelism is $\pi/2$ for all lengths.* (See TP 16 Notes, Claim 1)
> *6. The angle of parallelism is $\pi/2$ for one length.* (See TP 22)[1]
> *7. Similar, non-congruent figures exist.* (See TP 20 Notes, Claim 6)
> *8. A rectangle exists.* (See TP 20 Notes, Claim 7)
> *9. Legendre's assumption.* (See TP 19 Notes, Claim 2)
> *10. Two parallels have a common perpendicular.* (See TP 22)

Euclid could have made any of these assumptions his fifth postulate, and deduced the same body of results that comprise the *Elements*. There are, of course, still other equivalent statements, but we shall not dwell upon them here; it is time to bid farewell to the parallel postulate altogether.

Heretofore, Lobachevski has developed only *neutral* theorems. A rigorous demonstration of the parallel postulate would instantly reduce them to an eccentric sequence of trivialities with unnecessarily difficult proofs, but it would not divest them of their validity. Henceforth, however, the safety net of neutrality will be absent: a proof of the parallel postulate would render everything that follows in Lobachevski's work not merely trivial, but actually false. It is at this point that the Lobachevskian heresy begins. No longer content to avoid the parallel postulate, he shall openly deny it and develop the consequences. In doing so, he must forego rectangles and similar triangles, accept that the angle of parallelism is acute for every length, accept that the sum of the angles in every triangle falls short of 180°, and, on a personal level, accept the scorn and condescending pity of his contemporaries.

Lobachevski derived his heterodox faith in the logical consistency of imaginary geometry largely from his ability to develop a consistent set of trigonometric formulae under the assumption

---

[1] If such a length exists, then by TP 22, all triangles have angle sum π, hence the parallel postulate holds. The converse is obvious.

that the parallel postulate was false. Since Euclidean trigonometry is built upon consequences of the parallel postulate[2], the reader may appreciate what a difficult and remarkable achievement this was. Naturally, he had no choice but to found non-Euclidean trigonometry upon a different basis altogether. The geometric creativity he displays along the way is breathtaking, as we shall witness in the latter portions of the *Theory of Parallels*. The resulting trigonometric formulae make the *analytic* exploration of imaginary geometry possible, quickly yielding further results that testify, if not to the consistency of the new geometry, then at least to its beauty.

---

[2] Namely, it is built upon the theory of similar triangles, without which one cannot define the trigonometric functions as side-ratios.

Theory of Parallels 23

For any given angle α, there is a line p such that $\Pi(p) = \alpha$.

Let AB and AC be two straight lines forming an acute angle α at their point of intersection A. From an arbitrary point B' on AB, drop a perpendicular $B'A'$ to AC. Make $A'A'' = AA'$, and erect a perpendicular $A''B''$ upon A''; repeat this construction until reaching a perpendicular CD that fails to meet AB. This must occur, for if the sum of the three angles equals $\pi - a$ in triangle $\triangle AA'B'$, then it equals $\pi - 2a$ in triangle $\triangle AB'A''$, and is less than $\pi - 2a$ in $\triangle AA''B''$

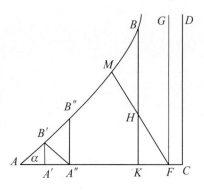

(TP 20); if the construction could be repeated indefinitely, the sum would eventually become negative, thereby demonstrating the impossibility of the perpetual construction of such triangles.

Beginning with the initial segment AA', the repeated construction consists of doubling the length of the segment and erecting a perpendicular at its new endpoint. In Euclidean geometry, each perpendicular would meet AB at an angle of $(\pi/2 - \alpha)$, producing an endless sequence of similar right triangles. In imaginary geometry, this cannot happen, as there are no similar triangles. After Lambert's results on area and angle defect (discussed in the notes to TP 20) we should not be surprised to learn that each successive perpendicular meets AB at a smaller angle than does its predecessor. However, Lobachevski asserts something stranger still: the repeated construction will eventually produce a perpendicular that actually fails to intersect AB.

His remarkably simple proof relies upon the additivity of defect. Because of this additivity, the defect of each successive right triangle must be at least twice that of its predecessor. (Proof: By Claim 4 of the TP 20 notes, $defect(\triangle AA''B'') > defect(\triangle AA''B') = [defect(\triangle AA'B') + defect(\triangle A''A'B')] = 2defect(\triangle AA'B')$, where the last equality holds because $\triangle A''A'B' \cong \triangle AA'B'$ by SAS.) Therefore, if the construction could be continued forever, yielding a new right triangle at each step, then we could repeat it sufficiently many times to make the defect of the resulting triangle exceed π. This is obviously impossible, so the process must eventually cease to yield triangles. That is, the perpendiculars erected past a certain point must fail to meet the line AB.

The perpendicular CD itself might have the property that all other perpendiculars closer to A cut AB. At any rate, there is a perpendicular FG at the transition from the cutting-perpendiculars to the non-cutting-perpendiculars that does have this property. Draw any line FH making an acute angle with FG and lying on the same side of it as point A. From any point H of FH, drop a perpendicular HK to AC; its extension must intersect AB at some point; say, at B. In this way, the construction yields a triangle $\triangle AKB$, into which the line FH enters and must, consequently, meet the hypotenuse AB at some point M. Since the angle $\angle GFH$ is arbitrary and can be chosen as small as one wishes, FG is parallel to AB, and $AF = p$. (TP 16 and 18).

Lobachevski's assertion that the boundary-perpendicular FG exists should remind the reader of a similar assertion concerning parallels in TP 16. Lobachevski takes its existence for granted, but we can rigorously prove its existence by appealing to properties of the real numbers, as follows. Each point on ray AC has a nonnegative real number associated with it (its distance from A), and vice versa. Consider the set of reals that correspond to points whose perpendiculars cut AB. Since, as we have seen, the perpendiculars eventually cease to cut AB, this set of real numbers is bounded above. Hence, it has a least upper bound. This least upper bound corresponds to a point on AC, and it is not hard to show that the perpendicular erected there is the boundary-perpendicular FG that we desire.

The rest of Lobachevski's argument in this paragraph is straightforward. He requires, but does not cite, TP 2 and TP 3 to secure the existence of M. We know that $FG\|AB$, because these lines satisfy the criteria in the definition of parallelism (TP 16): they do not meet, but every ray FH that enters $\angle GFA$ does meet AB. Thus, by the symmetry of parallelism (TP 18), it follows that $AB\|FG$. Hence, α is the angle of parallelism for the line segment AF. Letting p be the length of AF, we write $\Pi(p) = \alpha$. Since α was an arbitrary acute angle, this demonstrates that every acute angle occurs as an angle of parallelism for some length, as was to be shown.

Further Notes on the Π-Function.

It is easy to see that with the decrease of p, the angle α increases, approaching the value $\pi/2$ for $p = 0$; with the increase of p, the angle α decreases, approaching ever nearer to zero for $p = \infty$.

Here, we prove a few basic facts about the Π-function.

Claim 1. Π is a decreasing function. (That is, if $q < p$, then $\Pi(q) > \Pi(p)$.)

Proof. Let AF be a line segment of length p; let $FG\perp AF$, and let $AB\|FG$, as in the figure. By definition, $\angle BAF = \Pi(p)$. As was demonstrated above, any perpendicular erected in the interior of the line segment AF will intersect AB. In particular, if we erect a perpendicular upon the point S such that $AS = q$, it will meet AB at some point T. We saw in TP 16 that all lines emanating from point A fall into two classes with respect to the line ST: the class of cutting-lines, and the class of non-cutting-lines. Since line

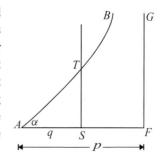

AT, which makes an angle of $\Pi(p)$ with AS, is a cutting-line, we know that the boundary-line separating the two classes (i.e. the parallel to ST through A) makes a greater angle with AS than $\Pi(p)$. That is, $\Pi(q) > \Pi(p)$, as claimed.   ■

This verifies that lengthening a line segment shrinks its angle of parallelism, while diminishing a segment increases its associated angle. Since *every* acute angle occurs as an angle of parallelism, we know that we may force the angle of parallelism as close to the Euclidean value of $\pi/2$ as we wish by taking a sufficiently small line segment. Similarly, with a sufficiently long segment, the angle of parallelism can be brought as close to 0 as we wish. This verifies the limiting behavior noted by Lobachevski.

Corollary. Π *is a continuous function.*

Sketch of Proof. Π is decreasing on its domain $(0,\infty)$ and assumes all values in its range $(0, \pi/2)$. (For a proof that continuity follows from these conditions, see any real analysis text, for example Bressoud, p. 100).   ■

Extending the Π-Function's Domain

Since we are completely free to choose the angle that shall be assigned to the symbol $\Pi(p)$ when p is a negative number, we shall adopt the convention that $\Pi(p) + \Pi(-p) = \pi$, an equation which gives the symbol a meaning for all values of p, positive as well as negative, and for $p = 0$.

Since the definition of $\Pi(p)$ (TP 16) presumes that p is a length, the domain of Π is initially restricted to positive values of p. However, nothing prevents us from assigning a meaning to $\Pi(p)$ when p is a negative number. Lobachevski defines it to be a shorthand notation for the *supplement* of an angle of parallelism. (for example, if $\Pi(p) = 77°$, then $\Pi(-p) = 103°$.) This will prove convenient in the later propositions of the *Theory of Parallels*. (We will not see this notation until TP 34.)

Lobachevski's extension changes Π's domain from the positive reals to all real numbers, and its range from $(0,\pi/2)$ to $(0,\pi)$. It preserves the continuity of Π. In particular, since $\pi = \Pi(0) + \Pi(-0) = 2\Pi(0)$, we have $\Pi(0) = \pi/2$, just as Π's limiting behavior at 0 would have us expect.

Although our quantitative understanding of the Π-function is still dim, we are beginning to obtain a qualitative image of its behavior. A rough, tentative sketch of the Π-function's graph, based on the limited information we possess would depict a monotone decreasing function that is bounded above and below by a pair of asymptotes separated from one another by a distance of π. Thus, it would bear some resemblance to the graph of $y = -\arctan(x) + \pi/2$.

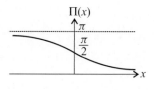

Intrinsic Measurements

In both Euclidean and imaginary geometry, we have an *intrinsic* unit of angle measurement: the right angle. Euclid's fourth postulate asserts that all right angles are equal, and his 11th proposition explicitly describes how to construct a right angle of your very own. This being the case, there is no need to keep a distinguished right angle in a pressurized chamber in the Bureau of Standards; right angles are built into the very fabric of geometry.

In contrast, Euclidean geometry has no intrinsic unit of length. We can relate our various, conventional units of length to one another—one mile equals 5280 feet, for example—but we cannot construct any of them directly from the axioms.

Astonishingly, imaginary geometry *does* admit intrinsic measures of length. TP 23 associates each acute angle with a unique length, and vice-versa. This intertwining of length (size) with angle (shape) makes it easy to parlay intrinsic measures of the latter into intrinsic measures of the former; we can define our standard unit of length to be, for example, the unique length whose angle of parallelism is $\pi/4$. Since the angle $\pi/4$, half of a right angle, is a concept intrinsic to geometry, the unique length associated with it (an imaginary geometry) is also an intrinsic measure.

Lambert was the first to notice the possibility of measuring lengths intrinsically in imaginary geometry. His reaction to this prospect is best conveyed by his own words.

> "This consequence is somewhat surprising, which inclines one to want the third hypothesis to be true! However, this advantage notwithstanding, I still do not want it, because innumerable other inconveniences would thereby come about. Trigonometric tables would have to be infinitely extended; the similarity and proportionality of figures would entirely lapse; no figure could be presented except in its absolute size; Astronomy would be an evil task; etc.
>
> But these are *argumenta ab amore et invidia ducta*[1], which Geometry, like all the sciences, must leave entirely on one side. I therefore return to the third hypothesis. . ."[2]

The particular choice of unit is a mere detail; the fact that we *can* choose an absolute unit is the great surprise. Whether we could actually behold such a unit is a question of a different nature. If our universe is described by imaginary geometry, then determining the size of an absolute unit of length might be impossible in practice. Compared to the vastness of interstellar space, any distance with which we have any experience is essentially infinitesimal. Since the angle of parallelism of every terrestrial length *looks like* $\pi/2$, finding a line segment whose angle of parallelism deviates perceptibly from this value might require us to examine segments whose lengths exceed the diameter of our galaxy. This is worth keeping in mind when looking at the figures in Lobachevski's text; although they fill but a few square inches on the page, they might represent geometric configurations occurring only on an astronomical scale.

Accordingly, Ferdinand Karl Schweikart (1780–1859) referred to the new geometry as *Astral Geometry* in a tantalizing fragment he sent to Gauss in 1818.

Schweikart

Schweikart attended mathematics lectures while studying law at Marburg in his late teens. Although law became his profession, he maintained a strong interest in mathematics, and in 1807, Schweikart published *Die Theorie der Parallellinien nebst dem Vorschlage ihrer Verbannung aus der Geometrie* ("The Theory of Parallel Lines Including a Proposal for its Banishment from

---

[1] Arguments drawn from love and hate.

[2] Fauvel and Gray, p. 518

Geometry"). Despite its promising title, the book does not proceed in Lobachevskian fashion; it simply develops Euclid's theory of parallels along lines that are slightly different from, but ultimately equivalent to, Euclid's own. It was Schweikart's only published mathematical work. Had it been his only contribution to geometry, his name would have been lost to history long ago.

By 1818, Schweikart's ideas about geometry had changed radically. Our primary piece of documentary evidence attesting to them is a brief note that he sent to Gauss through a mutual acquaintance, Christian Ludwig Gerling, an astronomer and a former student of Gauss. This note reads, in its entirety, as follows[3]:

Marburg, December 1818

There are two kinds of geometry—a geometry in the strict sense—the Euclidean; and an astral geometry.

Triangles in the latter have the property that the sum of their three angles is not equal to two right angles.

This being assumed, we can prove rigorously:

a) that the sum of the three angles of the triangle is less than two right angles;
b) that the sum becomes ever less, the greater the area of the triangle;
c) that the altitude of an isosceles right-angled triangle continually grows, as the sides increase, but it can never become greater than a certain length, which I call the *Constant*.

Squares have, therefore, the following form:

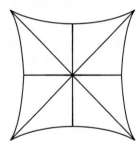

If this Constant were for us the Radius of the Earth, (so that every line drawn in the universe from one fixed star to another, distant $90°$ from the first, would be a tangent to the surface of the earth), it would be infinitely great in comparison with the spaces which occur in daily life.

The Euclidean geometry holds only on the assumption that the Constant is infinite. Only in this case is it true that the three angles of every triangle are equal to two right angles: and this can easily be proved, as soon as we admit that the Constant is infinite.

Schweikart.

Schweikart's surprising insight may have stemmed from the fact that he read Lambert's work on parallels, which he mentions in his own book of 1807. Schweikart's first two statements about astral geometry, (a) and (b), were known to Lambert, but the third (c) is original. It is unclear

---

[3] This translation is taken from Bonola, p.76. Schweikart's original German version is reproduced in Gauss, pp. 180–181.

just how much further Schweikart penetrated into non-Euclidean geometry than Lambert. Be that as it may, there is no question that he took an important psychological step forward from his predecessors.

Schweikart's assertion that there are two geometries is striking; his seems to be the first written statement to this effect. Certainly, it indicates that Schweikart had a vastly different conception of mathematics than did Saccheri. The contrast between these two is particularly apt since we might reasonably consider Schweikart to be Saccheri's geometrical grandson: Saccheri seems to have inspired Lambert's work, which in turn inspired Schweikart.

Saccheri never doubted that Euclidean geometry was the only geometry. Indeed, his conviction was so strong that after dozens of pages of brilliant, closely argued reasoning, it drove him to avert his eyes and permit a logically suspect argument to enter his work for purely political reasons: it purported to "prove" that which his conviction told him must be true. Lambert, Age of Reason scientist as he was, was better able to divorce his convictions from his investigations (exemplified by his comments on *argumenta ab amore et invidia ducta* quoted above.) He never fooled himself into believing that he had found a contradiction, but as a result, he apparently considered his researches disappointingly inconclusive, and hence not worth publishing. Schweikart's simple but profound step was to acknowledge the strange theorems unearthed by his predecessors as aspects of an alternate, logically viable geometry. His suggestion that these theorems might actually apply to the physical universe is remarkable.

Schweikart's *Constant* is an intrinsic unit of length, built into the fabric of astral geometry. In fact, it is another characterization of the length whose angle of parallelism is $\pi/4$.[4] In a March 1819 letter to Gerling, Gauss expressed his pleasure at Schweikart's memorandum and offered some characteristic Gaussian praise. ("It could almost have been written by myself.") He entreated Gerling to congratulate Schweikart, adding that, "I have extended the Astral Geometry so far that I can fully solve all its problems as soon as the constant C is given." As an example, Gauss mentioned that the maximum area of a triangle in the new geometry is precisely $\pi C^2/[\log(1 + \sqrt{2})]$.[2]

Although Schweikart never published an account of astral geometry, he had one more role to play in the subject's history; he encouraged his nephew, Franz Adolph Taurinus (1794–1874) to study the subject, bringing another important figure into the story of pre-Lobachevski non-Euclidean geometry. We shall have more to say about Taurinus' work later, but two facts pertinent to the present discussion of absolute measures are worth mentioning here.

Taurinus believed that the new geometry was logically consistent, but unlike his uncle, he never believed that it might be applicable to reality. His objections stemmed from the fact that non-Euclidean geometry admitted an absolute measure of length.

Taurinus' studies led him into a correspondence with Gauss. In a letter to Taurinus dated November 8, 1824, Gauss wrote that:

"All my efforts to discover a contradiction, an inconsistency, in this non-Euclidean geometry, have been without success, and the one thing in it which is opposed to

---

[4] *Sketch of Proof.* Let $\triangle XYZ$ be an isosceles triangle with a right angle at X. The altitude XW bisects the right angle, so $\angle WXZ = \pi/4$. Thus, since ray XZ cuts WZ, we know that $\Pi(XW) < \pi/4$. If C represents the length with angle of parallelism $\pi/4$, it follows that the length of the altitude XW is less than C. Hence, the least upper bound of all possible altitudes—Schweikart's Constant—is less than or equal to C. In fact, since right isosceles triangles with altitudes arbitrarily close to C exist (this is easy to demonstrate), Schweikart's constant must be *equal* to C, as claimed.

our conceptions is that, if it were true, there must exist in space a linear magnitude, *determined for itself* (but unknown to us). But it seems to me that we know, despite the say-nothing word-wisdom of the metaphysicians, too little, or too nearly nothing at all, about the true nature of space, to consider as *absolutely impossible* that which appears to us unnatural. If this non-Euclidean geometry were true, and it were possible to compare that constant with such magnitudes as we encounter in our measurements on the earth and in the heavens, it could then be determined *a posteriori*. Consequently in jest I have sometimes expressed the wish that the Euclidean geometry were not true, since then we would have *a priori* an absolute standard of measure."[5]

He had expressed a similar sentiment as early as 1816, in a letter to Gerling:

"It seems paradoxical but there could be a constant straight line given as if *a priori*, but I do not find in this any contradiction. In fact, it would be desirable that Euclidean geometry were not true, for we would then have a universal measure *a priori*. One could use the side of an equilateral triangle with angle = $59°59'59''$,9999 as a unit of length."[6]

---

[5] Wolfe, p. 47

[6] Rosenfeld, p. 215

Theory of Parallels 24

The farther parallel lines are extended in the direction of their parallelism, the more they approach one another.

To prove this proposition, Lobachevski shows that if $CG\|AB$, then G is closer to AB than C is. This is easy to miss on a first reading, since an auxiliary construction dominates the proof; Lobachevski does not even mention the parallel CG until the penultimate sentence of his proof.

Upon the line AB, erect two perpendiculars $AC = BD$, and join their endpoints C and D with a straight line. The resulting quadrilateral $CABD$ will have right angles at A and B, but acute angles at C and D (TP 22[1]). These acute angles are equal to one another; one can easily convince oneself of this by imagining laying the

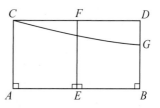

quadrilateral upon itself in such a way that the line BD lies upon AC, and AC lies upon BD. Bisect AB. From the midpoint E, erect the line EF perpendicular to AB; it will be perpendicular to CD as well, since the quadrilaterals $CAEF$ and $FEBD$ coincide when one is laid on top of the other in such a way that FE remains in the same place.

Once we know that the summit angles $\angle ACD$ and $\angle BDC$ are equal, their acuteness follows easily: were they obtuse, then $CABD$'s angle sum would exceed 2π, with the result that at least one of the two triangles formed by drawing the diagonal AD would violate the Saccheri-Legendre theorem (TP 19); were they right, then $CABD$ would be a rectangle—an impossible figure in imaginary geometry (TP 20 Notes, Claim 7).

Lobachevski uses superposition to demonstrate that $\angle ACD = \angle BDC$ and that $EF\perp CD$. Although open to Hilbertian criticism, such demonstrations are not lacking in value; at the very least, they strongly suggest the truth of the statements they purport to prove.[2] Those who prefer iron-clad proofs that $\angle ACD = \angle BDC$ and $EF\perp CD$ can find them in the opening pages of Saccheri's *Euclides Vindicatus*, to which we now turn.

---

[1] This refers to Lobachevski's declaration at the end of TP 22 that he would work in imaginary geometry from that point forward. Had he carried out this construction earlier, he would not have been able to deduce that the angles at C and D were acute; in neutral geometry, they could be either acute or right.

[2] Here is an even quicker intuitive proof: by construction, $CABD$ is symmetric about the perpendicular bisector of AB Hence, this bisector must be perpendicular to CD (lest it break the symmetry), and the angles at C and D must be equal.

A Sample of Saccheri

The quadrilateral *CABD* that Lobachevski constructs in TP 24 is an example of what is now called a *Saccheri quadrilateral*. (As discussed in the notes to TP 19, a Saccheri quadrilateral is formed by erecting equal perpendiculars upon a line segment's extremities, and joining their endpoints.) I have reproduced Saccheri's first two propositions below; these serve not only to demonstrate Saccheri's style, but also to supplement Lobachevski's superposition arguments with additional proofs of a style to which no one can object.

Preposition I. *If two equal lines AC, BD, form equal angles with the line AB: I say that the angles at CD will equal one another.*

Proof. Join *AD, CB*. Then consider the triangles *CAB, DBA*. It follows (Euclid I.4 [SAS-criterion]) that the sides *CB, AD* will be equal. Then consider the triangles *ACD, BDC*. It follows (Euclid I.8 [SSS-criterion]) that the angles *ACD, BDC* will be equal.   Q.E.D.   ■

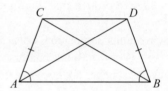

Preposition II. *Retaining the same quadrilateral ABCD, bisect the sides AB, CD at the points M and H. I say the angles at MH will be right.*

Proof. Join *AH, BH*, and likewise, *CM, DM*. Since the angles at *A* and *B* in this quadrilateral are presumed equal, and (by the preceding proposition) the angles at *C* and *D* are equal as well, it follows from Euclid I.4 (noting the equality of the sides) that in the triangles *CAM* and *DBM*, the sides *CM* and *DM* are equal; similarly, in triangles *ACH* and *BDH*, the sides *AH* and *BH* are equal. Consequently, by comparing first the triangles *CHM* and *DHM*, and then the triangles *AMH* and *BMH*, it follows (from Euclid I.8) that we have equal, and therefore right, angles at the points *M* and *H*.   Q.E.D.[3]   ■

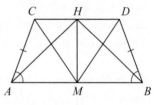

Lobachevski's claim that $\angle ACD = \angle BDC$ follows directly from Saccheri's first proposition, while his claim that *AB*'s perpendicular bisector meets *CD* at right angles follows from Saccheri's second proposition.

Note that in these first two propositions, Saccheri places no restriction on the base angles, other than their equality. Beginning in his third proposition, he restricts his attention to the quadrilaterals now named after him: those in which the base angles are both *right* angles.

Thabit ibn Qurra, Omar Khayyám, and the Politics of Naming

> "The fact is that every writer *creates* his own precursors.
> His work modifies our conception of the past . . ."
> —Jorge Luis Borges, "Kafka and his Precursors"

Saccheri's first two propositions were actually established long before Saccheri. The first dates back at least to Thabit ibn Qurra (836–901), and both propositions appear in the work of

---

[3] Saccheri, pp. 18–21.

Omar Khayyám (1045–1130). Under assumptions equivalent to the parallel postulate, both men proved that Saccheri quadrilaterals must be rectangles, and subsequently deduced the postulate from this fact. Like Saccheri, Khayyám established the HRA by proving the HOA and HAA untenable. This fact has led some writers (including Rosenfeld) to rechristen the relevant figure the *Khayyám-Saccheri quadrilateral*.

Lest accusations of Eurocentrism rain down upon me, I hasten to defend my retention of the traditional Khayyám-free designation. The quadrilateral is not important *per se*; rather, it is important because it acted as a window upon non-Euclidean geometry for an age that never suspected that a geometry other than Euclid's might exist.[4] In Saccheri's work, the window remained open for over thirty propositions, allowing him ample time to observe and describe the world he hoped to dispel. Although Khayyám did see the window, he opened and shut it in the very same proposition. If we sense something special in Khayyám's use of the quadrilateral today, it is only because we know the much more profound use to which Saccheri would put the same figure 700 years later.

Readers interested in the work of Khayyám and ibn Qurra on the parallel postulate can find detailed descriptions and extracts of it in Rosenfeld's book.[5]

Consequently, the line CD cannot be parallel with AB. On the contrary, the line from point C that *is* parallel to AB, which we shall call CG, must incline toward AB (TP 16), cutting from the perpendicular BD a part $BG <$ CA. Since C is an arbitrary point of the line CG, it follows that the farther CG is extended, the nearer it approaches AB.

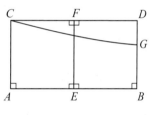

We may partition the set of rays through C into two classes: those that cut AB, and those that do not cut AB (TP 16). The boundary between the two classes is, of course, the unique parallel to AB that passes through C. Into which category does CD fall? Since it shares a common perpendicular with AB, it can be neither a cutting ray (TP 4), nor the parallel (TP 22). Hence, it is an ordinary undistinguished non-cutting ray. Accordingly, if we call the parallel CG, it follows that CG must enter angle $\angle ACD$. Naturally, it will meet BD at some point G. (Proof: CG cuts AD by the crossbar theorem applied to $\triangle ACD$; it then cuts BD by Pasch's axiom on $\triangle ADB$.) Since $BG < BD = AC$, the point G is closer to AB than the point C is, which was to be shown.

Thus, parallel lines draw ever closer to one another in their direction of parallelism, as claimed.

It is not yet clear whether parallel lines approach one another *asymptotically*, or whether the distance between parallels always remains greater than some finite positive value. This question will be settled in TP 33.

---

[4] Microsoft Word's grammar check objects to the phrase "a geometry," so perhaps this age is still with us.

[5] Rosenfeld pp. 49–56 (ibn Qurra), pp. 64–71 (Khayyám).

Theory of Parallels 25

Two straight lines parallel to a third line are parallel to one another.

Transitivity of Parallelism

Whereas the classical definition of parallelism is transitive only in the presence of the parallel postulate (see "A Deeper Definition" in the notes to TP 16), Lobachevski proves here that his new notion of parallelism is transitive in imaginary geometry as well.

His proof falls into two parts: first, he establishes transitivity in the plane; then, he does the same in space. The latter part, his first foray into three-dimensional imaginary geometry, initiates a sequence of results in solid geometry (culminating in TP 28). His desire to place these three-dimensional results together is responsible for this relatively late proof of transitivity: he could have presented it after TP 20.[1]

First Case: Transitivity in the Plane

We shall first assume that the three lines AB, CD, and EF lie in one plane.

Given three lines in a plane, two of which are parallel to the third, one of the three must lie between the other two.[2] We shall call it the "middle line"; the others we shall call "outer lines." Lobachevski's proof of transitivity in the plane is broken into two subcases. In the first subcase, the "third line" (the one to which the others are parallel) is an outer line; in the second subcase, it is the middle line.

middle line

Euclid himself demonstrated the transitivity of parallelism in Euclidean geometry (*Elements*, I.30, XI.9), so Lobachevski needs only to establish transitivity in imaginary geometry. Accordingly, any angle of parallelism that occurs in his proof will be acute. Both subcases are easy to follow, and require no illumination on my part.

---

[1] Lobachevski's two references to TP 22 in the proof of TP 25 are not to the proposition itself. Rather, they refer to the remark made after the proof of that proposition, that the angle of parallelism is either always acute or always right. Lobachevski could have noted this dichotomy earlier (after TP 20), but, for presumably dramatic purposes, he reserved it for the remarks immediately preceding his announcement in TP 22 of the "new geometric science, which I have named Imaginary geometry."

[2] This can be proved rigorously from Hilbert's axioms. Note that the lines must be parallel in the same direction.

If one of the outer lines, say AB, and the middle line, CD, are parallel to the remaining outer line, EF, then AB and CD will be parallel to one another. To prove this, drop a perpendicular AE from any point A of AB to EF; it will intersect CD at some point C (TP 5),[3] and the angle $\angle DCE$ will be acute (TP 22). Drop a perpendicular AG from A to CD; its foot G must fall on the side of C that forms an acute angle with AC (TP 9). Every line AH drawn from A into angle $\angle BAC$ must cut EF, the parallel to AB, at some point H, regardless of how small the angle $\angle BAH$ is taken. Consequently, the line CD, which enters the triangle $\triangle AEH$, must cut the line AH at some point K, since it is impossible for it to leave the triangle through EH. When AH is drawn from A into the angle $\angle CAG$, it must cut the extension of CD between C and G in the triangle $\angle CAG$. From the preceding argument, it follows that AB and CD are parallel (TP 16 and 18).

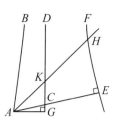

If, on the other hand, the two outer lines, AB and EF, are both parallel to the middle line CD, then every line AK drawn from A into the angle $\angle BAE$ will cut the line CD at some point K, regardless of how small the angle $\angle BAK$ is taken. Draw a line joining C to an arbitrary point L on the extension of AK. The line CL must cut EF at some point M, producing the triangle $\triangle MCE$. Since the extension of the line AL into the triangle $\triangle MCE$ can cut neither AC nor CM a second time, it must cut EF at some point H. Hence, AB and EF are mutually parallel.

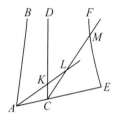

Having established transitivity in the plane, Lobachevski turns to space. As is its wont, the third dimension requires an entirely new proof. Adapting the two-dimensional proof is, unfortunately, impossible, since we cannot distinguish "outer" and "middle" lines in space; what looks like the middle line from one perspective will not from another. Before describing Lobachevski's proof of the spatial case, I must devote a few preliminary words to three-dimensional geometry.

Solid Geometry

The axioms of plane geometry were not designed to bear the weight of an extra dimension, so we must give some attention to the foundations when we move from the plane to space. In particular, we must secure some basic information about the behavior of planes in space, since Euclid's axioms tell us nothing about them. There are two ways to go about doing this.

The first way begins by formulating a precise definition of a plane. From the familiar axioms of plane geometry, one must then prove that any object satisfying the official definition of a plane actually behaves in a manner befitting of the name "plane." For example, one must prove, among other things, that there exists a unique plane passing through any three non-collinear points in space.

---

[3] Lobachevski's original text cites TP 3 here. I presume this was an editorial mistake.

Euclid attempted this procedure in Book XI of the *Elements*, but was not successful. His definition of a plane ("that surface which lines evenly with the straight lines on itself") is every bit as vague as his definition of a line. Unfortunately, the consequences of a non-defining definition are more serious in Book XI than they are in Book I. Euclid's attempted definition of a line is certainly an aesthetic failure—a superfluous utterance that mars an otherwise streamlined presentation. Yet however ugly it may be, it does no damage to the logical development of geometry, because Euclid never actually refers to it. Rather, he bases all his theorems about lines and rectilinear figures upon his axioms, which describe the properties of lines, and thus define them implicitly. Alas, since Euclid's axioms do *not* describe the properties of planes, his development of solid geometry is condemned to an awkward and illogical beginning. As even Thomas Heath acknowledges, "There is no doubt that the proofs of the first three propositions [in book XI] are unsatisfactory owing to the fact that Euclid is not able to make any use of his definition of a plane for the purpose of these proofs, and they really depend upon truths which can only be assumed as axiomatic."[4]

Euclid regains his usual composure in his fourth proposition, but his failure to provide an even moderately satisfactory foundation for solid geometry suggests that defining the term *plane* may not be the best strategy for moving from two to three dimensions. Given Euclid's failure, it is surprising to learn that his strategy is actually feasible (using a different definition than his, of course)! Carrying it to fruition, however, is delicate and difficult work, which would take us far afield. The most expedient path, which I shall follow here, is simply to accept *plane* as an undefined term, like *point* and *line*, and adopt a few extra axioms that describe the behavior of planes in space.

These "plane axioms" are:

1) If P and Q are points in a plane, then the entire line PQ lies in the plane.
2) Any three non-collinear points determine a unique plane.
3) Distinct planes either intersect in a line or do not meet at all.

These axioms appear, in slightly different form, in Hilbert's *Foundations of Geometry*, and correspond (roughly) to the first three propositions that appear in Book XI of Euclid's *Elements*. As I mentioned in the notes to TP 16, it is not my intention here to pursue a logically impeccable basis for geometry in the manner of Hilbert, but I would like to point out a notable feature of Hilbert's axioms for solid geometry: the jump from two to three dimensions does not necessitate any additional "betweenness" axioms. The spatial analogue of Pasch's axiom, for example, may be proved as a theorem from its two-dimensional counterpart and the plane axioms.

Because of the second plane axiom, we may employ the notation "plane ABC", without ambiguity, for the plane through three non-collinear points A, B, and C. If we wish to emphasize that some other point, say, D, also lies on plane ABC, we may refer to it as plane $ABCD$. In situations where it is clearly understood that we are referring to a plane, we may simply write ABC instead of *plane ABC*.

Lobachevski's definition of parallelism easily extends to space: *two lines in space are said to be parallel if they are coplanar and parallel in their common plane.* Given any line AB in space, and any point $P \notin AB$, it is easy to see that there is a unique line through P that is parallel

---

[4] Euclid, Vol. 3, p.272. Thomas Heath translated, edited, and annotated the standard English edition of the *Elements*.

to AB: if $PQ||AB$, then PQ and AB are coplanar, so any parallel PQ to AB must lie in plane ABP; by TP 16, there is one and only one such line.

Lobachevski's Lemma

Having disposed of these preliminaries, we turn to Lobachevski's proof of the transitivity of parallelism in space. He begins with a lemma: suppose that $AB||CD$; if a plane containing AB intersects a plane containing CD, then their line of intersection will be parallel to both AB and CD.

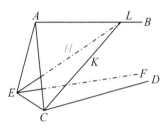

Suppose now that two parallels, AB and CD, lie in two planes whose line of intersection is EF. From an arbitrarily chosen point E of EF, drop a perpendicular EA to one of the parallels, say to AB. From the foot of this perpendicular, A, drop a new perpendicular, AC, to CD, the other parallel. Draw the line EC joining E and C, the endpoints of this perpendicular construction.

The angle $\angle BAC$ must be acute (TP 22), so the foot G of a perpendicular CG dropped from C to AB will fall on that side of AC in which the lines AB and CD are parallel. The line EC, together with any line EH that enters angle $\angle AEF$ (regardless of how slightly EH deviates from EF), determines a plane. This plane must cut the plane of the parallels AB and CD along some line CK. This line cuts AB somewhere— namely, at the very point L common to all three planes, through which the line EH necessarily passes as well. Thus, EF is parallel to AB. We can establish the parallelism of EF and CD similarly.

I have taken the liberty of changing the names of some of the points in this passage: the points I have called H, K, and L are *all* called H in Lobachevski's original.

Even after this change of notation, Lobachevski's proof remains awkward. To begin with, his sentence about line CG is irrelevant, a distraction that contributes nothing to the proof. (Accordingly, I have left CG off the figure, as its presence would add nothing but clutter.) Next, Lobachevski's immediate goal, to show that $EF||AB$, entails two steps: a proof that these lines do not meet, and a proof that any ray entering $\angle FEA$ must cut AB. He carries out the second step, but neglects the first entirely. Here is a more complete, and hopefully cleaner, version of his proof.

Claim 1 (Lobachevski's Lemma). Suppose that $AB||CD$; if a plane containing AB intersects a plane containing CD, then their line of intersection, EF, will be parallel to both AB and CD.

Proof. Draw the line segments EA, EC, and AC.

We shall prove that $EF||AB$. That is, we shall prove that these lines do not intersect, and that any ray entering $\angle FEA$ must cut AB. We shall do these one at a time.

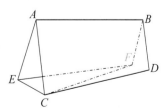

First, suppose by way of contradiction that AB and EF meet. Let X be their point of intersection. Since $X \in AB$ and AB lies on plane $ABCD$, it follows that X also lies on this plane.

Since X, C, and D all lie on this plane, planes XCD and $ABCD$ must be identical (by the second plane axiom). Similarly, $X \in EF$ implies that planes XCD and $EFCD$ are identical. Hence, planes $ABCD$ and $EFCD$ must be identical. This being the case, points A, B, C, D, E, F are coplanar. This, however, cannot be. (Proof: The intersecting planes in the statement of the lemma are clearly $ABEF$ and $CDEF$. If A, B, C, D, E, F were coplanar, then these planes would be identical, in which case they would not intersect in a line, contrary to hypothesis.) This contradiction shows that AB and EF do not intersect.

Next, we shall show that any ray entering $\angle FEA$ must cut AB. To this end, let ray EH enter $\angle FEA$. The lines EH and EC determine a plane: ECH. Because this plane cuts plane $ABCD$ at C, it must intersect $ABCD$ in some line through C (by the third plane axiom). Call it CK. Since CK enters $\angle DCA$ it must cut AB (because, by hypothesis, $CD \| AB$). Let L be the intersection of CK and AB.

Since $L \in CK$ and CK lies on plane ECH, it follows that L lies on plane ECH.

Since $L \in AB$ and AB lies on plane $ABEF$, it follows that L lies on plane $ABEF$.

Thus, L lies on the intersection of planes ECH and $ABEF$. That is, $L \in EH$. Having shown that ray EH cuts AB (at L), we conclude that $EF \| AB$, as claimed.

A similar argument shows that $EF \| CD$.

Hence, EF is parallel to both AB and CD, which was to be shown. ■

Transitivity at Last

Therefore, a line EF is parallel to one of a pair of parallels, AB and CD, if and only if EF is the intersection of two planes, each containing one of the parallels, AB and CD. Thus, two lines are parallel to one another if they are parallel to a third line, even if the lines do not all lie in one plane. This last sentence could also be expressed thus: the lines in which three planes intersect must all be parallel to one another if the parallelism of two of the lines is established.

In the first sentence of this passage, Lobachevski asserts the equivalence of two statements, but he does not actually bother to prove their equivalence. Clearly, the latter statement implies the former (by Lobachevski's lemma), but he gives no hint as to why the converse holds. I shall remedy this situation by proving a related theorem—fundamental in its own right—that will immediately establish Lobachevski's unproved claim, and with it, the transitivity of parallelism in space.

Claim 2. If two lines are parallel to a third, then the two lines must be coplanar.

Proof. Suppose that AB and CD are both parallel to EF. We shall show that AB and CD are coplanar.

The three points A, B, and C determine a plane. Since the planes ABC and $EFCD$ share point C, they must intersect in a line. Call it CK. By Lobachevski's Lemma (Claim 1), the intersection of plane $ABCK$ (which passes through AB) and plane $EFCDK$ (which passes through EF) must be parallel to both AB and EF. That is, $CK \| AB$ and $CK \| EF$.

CK is therefore a ray that lies in plane $EFCDK$, passes through C, and is parallel to EF. Since CD also meets this description, the uniqueness of parallels implies that $CK = CD$.

Since AB and CK were coplanar by construction, we conclude that AB and CD are coplanar, as claimed. ∎

Corollary (Lobachevski's unproved claim). *Suppose that $AB \| CD$. If EF is parallel to one of these lines, then EF is the intersection of a plane containing AB and a plane containing CD.*

Proof. Suppose, without loss of generality, that $EF \| AB$. Since EF and CD are both parallel to AB, we know that the lines EF and CD must be coplanar (Claim 2). Their common plane $EFCD$ contains CD; its intersection with plane $ABEF$ (which contains AB) is EF. ∎

Claim 3 (Transitivity at last). Parallelism is transitive in three-dimensional space.

Proof. Suppose that $AB \| CD$ and $AB \| EF$. By the preceding corollary, EF is the intersection of a plane containing AB and a plane containing CD. Thus, by Lobachevski's lemma (Claim 1), it follows that $EF \| CD$. Hence, parallelism is transitive in space, as claimed. ∎

Finally, Lobachevski observes that if three planes are arranged in such a way that two of their three lines of intersection are parallel, then all three must be parallel to one another. This follows directly from Lobachevski's lemma. The figure that  it suggests—an infinite triangular prism whose three edges are mutually parallel—will play an important role in much that follows. We shall meet it again in TP 28.

Bolyai's Proof

Bolyai demonstrates the transitivity of parallelism in §7 of his *Appendix*. His incisive proof of the transitivity of parallelism in space makes Lobachevski's proof look laboriously cobbled by comparison. Bolyai not only confirms the truth of the theorem, but also renders it intuitive. His use of motion is elegant, but it leaves his proof open to criticism of insufficient rigor. I shall retell Bolyai's proof in my own words, retaining his notation for the benefit of those readers who wish to compare it to Bolyai's terse original text.

Theorem. *Let BN, CP, and AM be lines that do not all lie in the same plane. If BN and CP are both parallel to AM, then BN and CP are parallel to one another.*

Proof. Let D be any point on AM. Rotate plane BCD about BC so that D moves along ray AM. This makes line BD rotate toward BN (in plane $AMBN$), and CD rotate toward CP (in plane $AMCP$). When plane BCD separates from AM, the cutting-lines BD and CD will cease to cut AM; at that moment, they will coincide with BN and CP, the unique parallels to AM through B and C respectively. Since BN and CP both lie on the rotating plane at the same moment, these 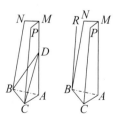 lines must be coplanar. Thus, we have demonstrated that if two lines are parallel to a third line, the two must be coplanar.

Next, we shall show that BN and CP are not merely coplanar, but parallel.

Let BR be the unique parallel to CP through B; we shall show that $BR = BN$.

Since BR and AM are both parallel to CP, the lines BR and AM must be coplanar, by the result we just established. Thus, BR lies in plane BAM, which is identical to plane $BNAM$. Of course, BR also lies in plane $BRCP$, which is identical to $BNCP$. (BN and CP are coplanar, so the plane $BNCP$ exists; it coincides with $BRCP$ by the second plane axiom.) Thus, BR is the intersection of planes $BNAM$ and $BNCP$. That is, $BR = BN$. Hence, $BN||CP$, as claimed. ■

Lobachevski, Bolyai, and Gauss all proved the transitivity of parallelism in the plane by examining two cases: when the "third line" lies between the other two, and when it does not. Gauss left no proof of the transitivity of parallelism in three dimensions.

Theory of Parallels 26

Antipodal spherical triangles have equal areas.

By antipodal triangles, I mean those triangles that are formed on opposite sides of a sphere when three planes through its center intersect it. It follows that antipodal triangles have their sides and angles in reverse order.

As Lobachevski notes in TP 12, a plane passed through a sphere produces a circle of intersection on the sphere's surface. The closer the plane comes to the sphere's center, the larger the circle of intersection will be; if it passes through the center itself, the resulting intersection is a *great circle*. On a globe, for example, lines of longitude and the equator are all examples of great circles, while the Tropic of Capricorn is not.

Diametrically opposed (*antipodal*) points on a sphere can be joined by many great circles; the North and South poles of the globe, for example, are connected by *all* lines of longitude. In contrast, any two *non*-antipodal points can be joined by one and only one great circle: the great circle lying in the unique plane determined by the sphere's center and the two non-antipodal points. The two points split "their" great circle into a pair of arcs, the shorter of which we shall call, conveniently if ungrammatically, *the* great circle arc that joins the points.

A *spherical triangle* consists of three points on a sphere and the great circle arcs that join them to one another. If A, B, and C are points on a sphere, and A', B', and C' are the points diametrically opposed to them, then the spherical triangles $\triangle ABC$ and $\triangle A'B'C'$ are said to be *antipodal triangles*. This unambiguous definition is both neater than, and equivalent to, Lobachevski's somewhat vague description of "triangles formed on opposite sides...", and thus is preferable to it on logical grounds.

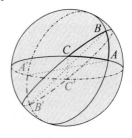

If, as in the figure, the vertices A, B, and C of a spherical triangle occur in counterclockwise order (when viewed from the perspective of a bug standing within the triangle on the surface of the sphere), the corresponding vertices A', B', and C' of its antipodal triangle will clearly occur in *clockwise* order (and vice versa). Obvious though this may be from the figure, one must define orientation to prove this formally. To do so would entail a lengthy digression that would gain us nothing but rigor. Since no new insight is to be won, and since we shall not need to discuss orientation ever again after this proposition, I shall not belabor this point.

Continuing therefore in the intuitive spirit of Lobachevski's treatment of orientation, we distinguish two different types of geometric congruence. Congruent figures on the same surface (plane or sphere) are said to be *directly* congruent if one of them can be slid along the surface until it coincides point for point with its mate. If this cannot be done, the congruent figures are *oppositely* congruent. Oppositely congruent figures (examples in the plane include alternate footprints, or the letter K and its mirror image) can be made to coincide by lifting one figure out of the surface, turning it over, and returning it to the surface in its reversed state.

When Euclid compared the areas or volumes of figures, he relied on a handful of unspoken assumptions about how area behaves. One such assumption was that congruent figures have equal area. Consequently, a proof that antipodal triangles are congruent (which we shall demonstrate shortly) would have satisfied Euclid that antipodal triangles have the same area. Note, however, that antipodal triangles are, in general, *oppositely* congruent, due to their mutually reversed orientations. Unlike Euclid, Lobachevski was unwilling to assume that oppositely congruent figures have the same area, preferring to prove this fact from the more modest assumption that *directly* congruent figures have equal area. He does not prove this explicitly in the *Theory of Parallels*, but the idea for such a proof is implicit in the present proposition's demonstration. In order to understand it, we shall first need to understand Lobachevski's simple criterion for deciding when two figures have the same area.

Lobachevski explains this criterion in the proposition's last line (which ought to have been the first line): "I adopt the following postulate: two figures on a surface are equal in area when they can be formed by joining or detaching equal parts." Keeping in mind that Lobachevski's "equal parts" are our "directly congruent figures," we can understand the postulate by looking at a couple of pictures. The two figures at right are not congruent, but according to Lobachevski, they have equal area, since they are formed from directly congruent pieces. Below and to the right, we see two trapezoids; these are directly congruent, and thus have the same area. After removing a triangle from the left trapezoid and then removing a directly congruent copy of it from the right trapezoid, Lobachevski's postulate tells us that the resulting "holey trapezoids" also have the same area as one another. Lobachevski's postulate is natural for any reasonable notion of area; Euclid makes the same assumptions (albeit implicitly) in the *Elements*, except in a stronger form, allowing himself the luxury of letting the "equal parts" be either directly or oppositely congruent to one another. We shall return to this theme after proving that antipodal triangles are congruent.

> The corresponding sides of antipodal triangles $\triangle ABC$ and $\triangle A'B'C'$ are equal: $AB = A'B'$, $BC = B'C'$, $CA = C'A'$. The corresponding angles are also equal: those at A, B, and C equal those at A', B', and C' respectively.

Before we can prove that the corresponding angles of antipodal triangles are equal, we must be clear about how to measure the angles in a spherical triangle. In general, we define the measure of the angle at which two curves intersect to be the measure of the angle between their tangent lines at their point of intersection. However, when we restrict our attention to measuring the angles between great circles on a sphere, there is an alternate, equivalent method that is often more convenient: the measure of the angle between two great circles is equal to the measure of the angle of intersection between the planes in which the circles lie. We shall soon prove that these two methods of measurement yield the same result, but first, we must describe how one actually measures the angle between two intersecting planes.

A Dihedral Digression

The angle formed by two planes at their line of intersection is called a *dihedral angle*.
We measure such an angle as follows. From an arbitrary point of a dihedral angle's
"hinge" (the line in which the two planes meet), erect two perpendiculars, one in
each plane. These perpendiculars are called *lines of slope* for the dihedral angle. We
define the dihedral angle's measure to be the measure of the plane angle between its
lines of slope. To dispel what appears to be an ambiguity in this definition, we must
prove that the angle between a dihedral angle's lines of slope does not depend upon the point
from which they emanate. The standard proof of this fact, which one may find in any old textbook
on solid geometry, relies upon the parallel postulate, and thus is insufficient for our purposes.
The following proof, however, is neutral, and therefore acceptable. I have taken it from D.M.Y.
Sommerville's 1914 text, *The Elements of Non-Euclidean Geometry*.

Claim 1. In a dihedral angle, the measure of the plane angle between the lines of slope is
independent of the point from which they emanate. (i.e. the measure of a dihedral angle is
well-defined.)

Proof. Let α and β be two planes forming a dihedral
angle, and let P and P' be arbitrary points on their line
of intersection. Draw lines of slope PA and $P'A'$ in α
such that $PA = P'A'$, and lines of slope PB and $P'B'$ in
β such that $PB = P'B'$. To establish the claim, we must
show that $\angle APB = \angle A'P'B'$. To do so, we shall prove
that $\triangle APB \cong \triangle A'P'B'$.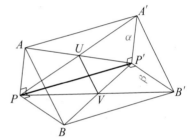

Let U be the point of intersection of PA' and $P'A$.

Note that $PU = P'U$. (Proof: $\triangle AP'P \cong \triangle A'PP'$ by SAS, so $\angle AP'P = \angle A'PP'$. These
equal angles are both in $\triangle PUP'$, so this triangle is isosceles, by Euclid I.6. That is, $PU = P'U$.)

Next, if we let V be the intersection of PB' and $P'B$, and apply the same argument to plane
β that we just used in plane α, we will find that $PV = P'V$.

We therefore know that $\triangle PUV \cong \triangle P'UV$, by SSS. Hence, $\angle UPV = \angle UP'V$.

Thus, $\triangle P'AB \cong \triangle PA'B'$ by SAS. ($PA' = P'A$ since $\triangle APB \cong \triangle A'P'B'$, as was shown
above. Similarly, $P'B = PB'$.)

Hence, $AB = A'B'$, from which it follows that $\triangle APB \cong \triangle A'P'B'$, by SSS.

Thus, $\angle APB = \angle A'P'B'$, which was to be shown. ∎

We can now verify our earlier claim that angles between great circles can be measured with
dihedral angles.

Claim 2. The measure of the angle between two great circles equals the measure of the dihedral
angle between the planes in which the great circles lie.

Proof. On a sphere with center O, let $\angle BAC$ be the angle
formed at A by two great circle arcs, AB and AC. By definition, the
measure of the spherical angle $\angle BAC$ equals the measure of the angle
between the tangent lines to arcs AB and AC at A. These tangent
lines lie in the planes OAB and OAC, which form a dihedral
angle with hinge OA. Since the tangents to the great circles are
perpendicular to radius OA, they constitute lines of slope for the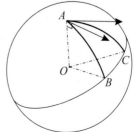

dihedral angle. Thus, the angle between the tangent lines measures not only the spherical angle $\angle BAC$, but also the dihedral angle. Since these last two angles are therefore equal, we may measure one with the other. That is, the angle between the great circles has the same measure as the dihedral angle formed by the planes in which the circles lie. ■

Antipodal Triangles are Congruent

> "But in regard to the story of the *antipodes*, that is, that there are men on the other side of the earth where the sun rises when it sets for us, who plant their footprints opposite ours, there is no logical ground for believing this."
> —St. Augustine, *The City of God Against the Pagans*.

In a plane, vertical angles are equal to one another. Applying this familiar result to the definition of dihedral angle measure, we can easily deduce that vertical dihedral angles are equal to one another. This fact, which Lobachevski notes in TP 6, will be used in the following simple proof.

Claim. Antipodal triangles are congruent.

Proof. Let $\triangle ABC$ and $\triangle A'B'C'$ be antipodal triangles on a sphere with center O.

The planes containing the great circles $ABA'B'$ and $ACA'C'$ form a dihedral angle, which measures the spherical angle $\angle A$; its vertical angle measures $\angle A'$. Since these vertical dihedral angles are equal, the spherical angles they measure are equal. That is, $\angle A = \angle A'$. Similarly, $\angle B = \angle B'$ and $\angle C = \angle C'$.

In any circle, equal central angles subtend equal arcs. Thus, in the great circle $ABA'B'$, whose center is O, the arcs AB and $A'B'$ must be equal, since they are subtended by the vertical (and hence equal) angles, $\angle AOB$ and $\angle A'OB'$. Similarly, $AC = A'C'$ and $BC = B'C'$.

Since all corresponding parts of the antipodal triangles are equal, the triangles are congruent (TP 15). ■

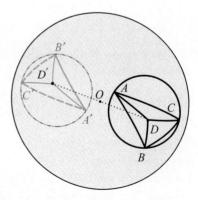

Consider the plane passing through the points A, B, and C. Drop a perpendicular to it from the center of the sphere, and extend this perpendicular in both directions; it will pierce the antipodal triangles in antipodal points, D and D'. The distances from D to the points A, B, and C, as measured along great circles of the sphere, must be equal, not only to one another (TP 12), but also to the distances $D'A'$, $D'B'$, and $D'C'$ on the antipodal triangle (TP 6). From this, it follows that the three isosceles triangles that surround D and comprise the spherical triangle $\triangle ABC$ are congruent to the corresponding isosceles triangles surrounding D' and comprising $\triangle A'B'C'$.

As a basis for determining when two figures on a surface are equal, I adopt the following postulate: two figures on a surface are equal in area when they can be formed by joining or detaching equal parts.

We have now seen that antipodal triangles are oppositely congruent. Essentially, Lobachevski would admit that antipodal triangles T and T' have the same area only after demonstrating that T can always be sliced into pieces, from which the antipodal triangle T' can subsequently be rebuilt, without having to turn any of the pieces over in the slicing and rebuilding process.

To understand why this is possible, consider the analogous situation in the Euclidean plane. The figure at right depicts oppositely congruent triangles; like any such pair, they are mirror images of one another. I have split each of them into three isosceles subtriangles by joining their vertices to their circumcenters.[1] It is clear that by detaching the pieces of the left triangle from one 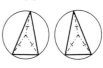 another, sliding them around the plane, and rejoining them, we can build a second copy of the right triangle from the pieces of the left triangle, without having to turn any of them over. Hence, the two triangles must have the same area (by Lobachevski's postulate). This elegant argument proceeds from the fact that *isosceles* triangles are directly congruent to their mirror images; they do not change their appearance if we turn them over. Thus, by breaking an arbitrary triangle into "nice" (i.e. isosceles) subtriangles, we can parlay a desirable property that occurs on a small scale (isosceles triangles have the same area as their mirror images) to a desirable large-scale property (all triangles have the same area as their mirror images).

Note that if the circumcenters do not lie within the triangles, we must adjust the above argument slightly. If the circumcenters lie *upon* the triangles, the adjustment is trivial—the triangles simply split into 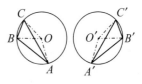 two isosceles subtriangles each, instead of three. If the circumcenters lie *outside* the triangles, a change of perspective will fix the argument: we view each triangle as the result of joining two isosceles triangles and subsequently detaching a third from their union. (For example, in the figure at right, we obtain $\triangle ABC$ by joining $\triangle AOB$ to $\triangle BOC$ and detaching $\triangle AOC$). Thus, Lobachevski's postulate for the equality of area, discussed above, implies the equality of area for these triangles.

Of course, Lobachevski's oppositely congruent triangles lie not in the Euclidean plane, but on the surface of a sphere in three-dimensional imaginary space. However, the ideas underlying his spherical proof are identical to those I have outlined in the preceding paragraphs. Consequently, I shall not discuss the mechanics of his proof in further detail. Rather, I wish to point out a curious phenomenon regarding circumcircles that will be important in later propositions.

Circumstantial Circumcircles

> "Weave a circle round him thrice,
> And close your eyes with holy dread..."
> — Samuel Taylor Coleridge, *Kubla Khan*

I specifically described the proof that triangles have the same area as their mirror images as taking place within the *Euclidean* plane because the ability to construct a triangle's circumcircle

---

[1] For every triangle in the Euclidean plane, there is a unique circle that circumscribes it (Euclid IV.5). This circle is the triangle's *circumcircle*; its center is the triangle's *circumcenter*. I shall discuss this in further detail in the notes to TP 29.

(Euclid IV.5) depends upon the parallel postulate! Indeed, the statement, "a circle may be circumscribed about any triangle" is equivalent to the parallel postulate. We shall prove this surprising fact in TP 29, and unfold its remarkable geometric consequences in subsequent propositions.

As the proof of TP 26 depends heavily upon our ability to circumscribe circles about arbitrary triangles, it is interesting to note the sources of our ability to do so in different geometric settings. In Euclidean geometry, the relevant source is the parallel postulate. In contrast, the fact that the intersection of a sphere and a plane is a circle is the key in spherical geometry: the vertices of any spherical triangle determine a plane, whose intersection with the sphere is the triangle's circumcircle.

Theory of Parallels 27

A trihedral angle equals half the sum of its dihedral angles minus a right angle.

Measuring Solid Angles

A solid angle is *dihedral* if it is bounded by two planes meeting at a line; *trihedral* if it is bounded by three planes at a point. Thus, a tetrahedron contains four trihedral angles, one at each vertex, and six dihedral angles, one at each edge. TP 27 relates the measure of a trihedral angle to the measures of the three dihedral angles at its edges. We know how to measure dihedral angles (see the TP 26 notes), but how does one measure a trihedral angle? We must answer this question before we can understand the statement of TP 27, much less prove it.

We tend to associate angle measurement with rotation. The measure of an ordinary angle in the plane, for example, indicates the amount of rotation required to bring one of its arms into coincidence with the other. Similarly, the measure of a dihedral angle indicates the amount of rotation required to bring one of its faces into coincidence with the other. The link between angle measure and rotation, however, ceases to exist when we work with trihedral angles (or more generally, when we work with *polyhedral* angles, formed by three or more planes meeting a point). Fortunately, we can articulate a general definition of polyhedral angle that agrees with our existing measures, but that is not based on rotation.

To motivate this definition, let us examine an alternate, protractor-free method for measuring ordinary angles in the plane. We begin by assigning a numerical value to the "full angle" (a $360°$ rotation). Any positive value (including 360) is permissible, but I shall set the full angle's value at 2π, so as to agree with Lobachevski. Then, to define the measure θ of an arbitrary angle in the plane, we proceed as follows. We draw an arbitrary circle (of circumference C) about the angle's vertex, and let s be the arc length of that part of it contained between the angle's arms. Clearly, the ratio $s:C = \theta:2\pi$ holds. Rewriting this as an equivalent formula, we obtain our definition of angle measure: $\theta = (s/C)2\pi$. That is, we define the measure of the angle to be 2π times the ratio of the subtended circular arc to the whole circle.

Naturally, making this definition fully rigorous would require a proof that the ratio $s:C$ is independent of the particular circle that we draw about the angle's vertex. This independence is essentially a consequence of the symmetry of circles together with the homogeneity of the plane (inasmuch as the plane "looks the same" in every direction[1]), but a proof requires some work. Since my concern, however, is with the fruits, rather than the roots, of this definition, I shall take

---

[1] Some prefer to use a separate term, *isotropy*, for this property of looking the same in every direction, reserving "homogeneity" or the property of looking the same at every point.

this objection as met, and proceed to generalize the definition so it that it covers solid angles (i.e. dihedral and polyhedral angles) as well.

Let 2π be the numerical value of the "full solid angle". We define the measure ϕ of an arbitrary solid angle as follows. We draw a sphere about its vertex (or in the case of a dihedral angle, about any point on its hinge). The angle will subtend a certain figure on the sphere (it will be a "lune" in the case of a dihedral angle, a spherical triangle in the case of a trihedral angle, etc.). This figure will cover a certain fraction of the sphere's surface. Multiply this fraction by 2π. We define the result to be the solid angle's measure.

For example, consider the dihedral angle formed by two perpendicular planes. When we center a sphere about a point on its hinge, the resulting subtended figure will comprise half of a hemisphere; that is, it will comprise $\frac{1}{4}$ of the entire sphere. Thus, the measure of the dihedral angle between the two perpendicular planes will be, according to our new definition, $\frac{1}{4}(2\pi) = \pi/2$, as expected.

One final note: Lobachevski follows a geometric tradition in which a genuine angle delimits a *convex* region of the plane (or space). Thus, the measure of every genuine angle (plane or solid) lies between 0 and π. This geometric convention, as opposed to the analytic convention, measures angles unambiguously. Angle measures will, moreover, lie *strictly* between 0 and π: a "flat angle" of measure 0 or π would not be considered a proper angle at all, since it would be a ray, a straight line, a half-plane, or a plane.

Now that we know what trihedral angles are, and how to measure them, we are almost capable of understanding the statement of TP 27. All that remains is to mention that for any trihedral angle, "its" dihedral angles are those formed at its three edges by the planes that meet at its vertex. The statement of TP 27 should now be comprehensible. Let us – at last – examine its proof.

The Size of Spherical Figures

Let $\triangle ABC$ be a spherical triangle, each of whose sides is less than half a great circle. Let A, B, and C denote the measures of its angles. Extending side AB to a great circle divides the sphere into two equal hemispheres. In the one containing $\triangle ABC$, extend the triangle's other two sides through C, denoting their second intersections with the great circle by A' and B'. In this way, the hemisphere is split into four triangles: $\triangle ABC$,

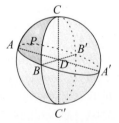

$\triangle ACB'$, $\triangle B'CA'$, and $\triangle A'CB$, whose sizes we shall denote by P, X, Y, and Z respectively.

Since the statement of TP 27 is concerned with trihedral angles, we naturally expect Lobachevski to begin his proof by specifying one. This is indeed what he does, although a reader today may not recognize it in the passage above. Nevertheless, Lobachevski is working with an arbitrary trihedral angle: he calls its vertex D; he calls the three planes that meet there ABD, ACD, and BCD. This naming, however, has occurred behind the curtain, as it were. When it rises, we, the audience, find Lobachevski *in medias res*, having already brought his trihedral angle out, named it parts, and placed a sphere about its vertex. We find him considering the figure that the trihedral angle subtends upon the sphere—a spherical triangle. Moreover, since his geometric convention for angles dictates that any genuine trihedral angle delimits a convex region of space, the spherical triangle's sides will each take up less than half of a great circle.

Now that we have recognized that Lobachevski's spherical triangle is the figure subtended by an arbitrary trihedral angle upon a sphere centered and its vertex, we can return to his words to see what he is trying to tell us about it. By extending its sides, he divides one hemisphere into four spherical triangles. He then refers to the "sizes" (*die Größe*) of these triangles. He defines (implicitly) the size of a spherical figure to be the measure of the central solid angle that subtends it. Thus, the full sphere's size is 2π (since this is the measure of the full solid angle), a hemisphere's size is π, and in general, if a figure takes up a certain fraction of the sphere's surface, its size is 2π times that fraction. Size is therefore directly proportional to area. The following simple lemma about the size of a spherical lune will prove helpful shortly.

Lemma. *The size of a spherical lune (a figure bounded by two distinct arcs that span the same pair of antipodal points) equals the measure of the angle at which its two arcs meet.* (See the figure.)

Proof. If θ is the angle between the arcs, then the lune clearly covers $\theta/(2\pi)$ of the sphere's total surface. Hence, its size is $(\theta/2\pi)(2\pi) = \theta$, as claimed. ∎

Clearly, $P + X = B$, and $P + Z = A$. Moreover, since the size Y of the spherical triangle $\triangle B'CA'$ equals that of its antipodal triangle $\triangle ABC'$ [TP 26], it follows that $P + Y = C$. Therefore, since $P + X + Y + Z = \pi$, we conclude that $P = \frac{1}{2}(A + B + C - \pi)$.

This swift, elegant proof amounts to little more than a threefold application of the lemma. The lemma gives $P + X = B$ and $P + Z = A$. Since antipodal triangles are congruent (TP 26 notes, Claim 3), they cover the same fraction of the sphere's surface, and thus have the same size. Hence, the lemma gives $P + Y = C$.

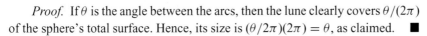

Summing the left and right-hand sides of the three equations yields

$$2P + (P + X + Y + Z) = A + B + C.$$

A glance at the figure reveals that the expression in parentheses represents the size of one hemisphere: π. Making this substitution and solving for P, we obtain

$$P = \frac{1}{2}(A + B + C - \pi).$$

By Claim 2 in the TP 26 notes, A, B, and C measure not only the angles of spherical triangle $\triangle ABC$, but also the dihedral angles between the three planes comprising the trihedral angle at D. Thus, we have proved the theorem.

TP 27 Rephrased as a Theorem about Spherical Triangles

Every trihedral angle that we shall meet in *The Theory of Parallels* will have its vertex at a sphere's center. Happily, we can rephrase TP 27 in a form specifically suited to this circumstance, since we may easily relate the trihedral angle and its three dihedral angles to features of the triangle it subtends upon the sphere's surface.

First, the *trihedral angle* equals the spherical triangle's *size*.

Second, the three dihedral angles are equal to the spherical triangle's three angles (TP 26 Notes, Claim 2). Thus, in particular, the *sum of the dihedral angles* equals the spherical triangle's *angle sum*.

Making these two replacements transforms Lobachevski's original statement of TP 27 ("*A trihedral angle equals half the sum of its dihedral angles minus a right angle*") into the following equivalent statement about spherical triangles:

$$\text{Size} = \tfrac{1}{2}(\text{Angle Sum}) - \pi/2.$$

Or equivalently,

$$\text{Size} = \tfrac{1}{2}(\text{Angle Sum} - \pi).$$

Finally, if we call the quantity in parentheses the *angular excess* of the spherical triangle (the amount by which its angle sum exceeds π), we obtain our desired relationship, which we express formally in the following restatement.

Claim 1. (TP 27 Rephrased). In any spherical triangle, the following relation holds:

$$\text{Size} = \tfrac{1}{2}(\text{Angular Excess}).$$

From this reformulation of TP 27, an important consequence follows.

Claim 2. In both Euclidean and imaginary geometry, every spherical triangle has angle sum greater than π.

Proof. Since a spherical triangle's size is a positive number, the preceding equation implies that its angular excess must also be positive. (Note that the equation holds in both geometries since TP 27 is a neutral result.) Thus, its angle sum must exceed π, as claimed. ■

A Theorem So Nice, He Proved It Twice

It is also possible to reach this conclusion by another method, based directly upon the postulate on equivalence of areas given above [in TP 26].

Curiously, Lobachevski gave two proofs of TP 27. His first proof, which we have just seen, resembles a sleight of hand magic trick; he diverts our attention from $\triangle ABC$ with three auxiliary triangles, and craftily extracts his formula while our eyes are elsewhere. If his first proof is magical, his second is economical. It requires less machinery, as it eschews theorem TP 26 altogether. Moreover, the first proof produces quite a bit of "waste" (the three auxiliary triangles); the second produces none: by repeatedly remolding the same material, it produces the theorem's conclusion from its premise in a strikingly direct fashion.

Like the first, the second proof begins with the spherical triangle $\triangle ABC$ that an arbitrary trihedral angle subtends. Lobachevski will show that the triangle's size (and hence the trihedral angle's measure) is equal to half its angular excess. To accomplish this, he cuts the triangle into pieces, and rearranges them to form a spherical Saccheri quadrilateral. He then dissects the quadrilateral as well, and from its pieces, he constructs a spherical lune, whose size can be shown to equal half the original triangle's angular excess. By Lobachevski's postulate on equality of areas (introduced in TP 26), the size of the original triangle must also equal half the original triangle's angular excess. Q.E.D.

Naturally, the devil is in the details.

Efficient though the second proof is, one cannot help but wonder why Lobachevski felt the need to include it. Despite its theoretical simplicity of means, it contains details that cry out for verifications of their own (only some of which Lobachevski provides), making it considerably longer than the first proof, at least on the printed page. Since it can be omitted without damaging the logical flow of the *Theory of Parallels*, I have confined my illumination to a mere partition of the proof into bite-sized pieces, for the sake of those readers who do wish to work through its details for themselves. Those who are satisfied with the first proof may safely skip the second one, and move directly to TP 28.

Here is the first chunk, in which Lobachevski shows how to construct a spherical Saccheri quadrilateral with the same area as the given spherical triangle. The construction is a familiar one, normally used (in the Euclidean plane) to prove that a triangle's area is half the product of its base and height. The three cases that Lobachevski considers here (*H* falls either within segment *DE*, upon one of the segment's endpoints, or outside the segment) arise in the Euclidean context as well.

In the spherical triangle $\triangle ABC$, bisect the sides AB and BC, and draw the great circle through D and E, their midpoints. Drop perpendiculars AF, BH, and CG upon this circle from A, B, and C. If H, the foot of the perpendicular dropped from B, falls between D and E, then the resulting right triangles 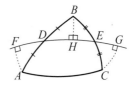 $\triangle BDH$ and $\triangle AFD$ will be congruent, as will $\triangle BHE$ and $\triangle EGC$ (TP 6 & 15).

From this, it follows that the area of triangle $\triangle ABC$ equals that of the quadrilateral $AFGC$.

If H coincides with E, only two equal right triangles will be produced, $\triangle AFD$ and $\triangle BDE$. Interchanging them establishes the equality of area of triangle $\triangle ABC$ and quadrilateral $AFGC$.

Finally, if H falls outside triangle $\triangle ABC$, the perpendicular CG must enter the triangle. We may then pass from triangle $\triangle ABC$ to quadrilateral $AFGC$ by adding triangle $\triangle FAD \cong \triangle DBH$ and then taking away triangle $\triangle CGE \cong \triangle EBH$.

In the next passage, Lobachevski establishes another property of the Saccheri quadrilateral: each of its remote angles equals half of the original triangle's angle sum.

Since the diagonal arcs AG and CF of the spherical quadrilateral $AFGC$ are equal to one another (TP 15), the triangles $\triangle FAC$ and $\triangle ACG$ are congruent to one another (TP 15), whence the angles $\angle FAC$ and $\angle ACG$ are equal to one another. Hence, in all the preceding cases, the sum of the three angles in the spherical triangle equals that of the two equal, non-right angles in the quadrilateral.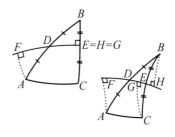

Therefore, given any spherical triangle whose angle sum is S, there is a quadrilateral with two right angles of the same area, each of whose other two angles equals $S/2$.

Next, he rearranges the Saccheri quadrilateral into a spherical lune.

Let $ABCD$ be such a quadrilateral, whose equal sides AB and DC are perpendicular to BC, and whose angles at A and D each equal $S/2$. Extend its sides AD and BC until they meet at E; extend AD beyond E to F, so that $EF = ED$, and then drop a perpendicular FG upon the 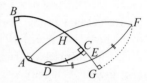 extension of BC. Bisect the arc BG, and join its midpoint H to A and F with great circle arcs.

Finally, Lobachevski shows that the lune's size is half of the original triangle's angular excess, completing his second proof.

The congruence of the triangles $\triangle EFG$ and $\triangle DCE$ (TP 15) implies that $FG = DC = AB$. The right triangles $\triangle ABH$ and $\triangle HGF$ are also congruent, since their corresponding arms are equal. From this it follows that the arcs AH and AF belong to the same great circle. Thus, the arc AHF is half a great circle, as is the arc $ADEF$. Since $\angle HFE = \angle HAD = S/2 - \angle BAH = S/2 - \angle HFG = S/2 - \angle HFE - \angle EFG = S/2 - \angle HAD - \pi + S/2$, we conclude that $\angle HFE = \frac{1}{2}(S - \pi)$. Equivalently, we have shown that $\frac{1}{2}(S - \pi)$ is the size of the spherical lune $AHFDA$, which in turn equals the size of the quadrilateral $ABCD$; this last equality is easy to see, since we may pass from one to the other by first adding the triangles $\triangle EFG$ and $\triangle BAH$, and then removing triangles that are congruent to them: $\triangle DCE$ and $\triangle HFG$.

Therefore, $\frac{1}{2}(S - \pi)$ is the size both of the quadrilateral $ABCD$, and of the spherical triangle, whose angle sum is S.

Theory of Parallels 28

If three planes intersect one another along parallel lines, the sum of the three resulting dihedral angles is equal to two right angles.

The Prism Theorem

This portentous result, which Jeremy Gray has named *the prism theorem*, says that if the edges of an infinitely long triangular prism are parallel to one another, then the three dihedral angles at those edges will add up to π. What makes this theorem remarkable is its neutrality. Its independence from the parallel postulate is surprising when one considers its resemblance to another theorem, Euclid I.32 (the sum of the angles in a triangle add up to π), which is actually equivalent to the postulate.

The prism theorem occupies a distinguished place in the structure of the *Theory of Parallels*. Thinking of the work as a drama in four acts, we might say that the first takes place in the plane, introduces the players and their concerns (Does the parallel postulate hold? What if it doesn't?), and ends halfway through TP 25. In the brief second act (TP 25–28), the setting shifts to three-dimensional space. The relevance of this shift is not immediately clear, but the dramatic entry of the prism theorem, just before the curtain falls for intermission, suggests that there is a hidden link between the first two acts after all. Act three (TP 29–34) slowly reveals this connection, and culminates with the construction of a surface called the *horosphere*. At the act's climax, the prism theorem returns to demonstrate that the horosphere, a surface in *imaginary* space, is endowed with an intrinsically *Euclidean* geometry. The consequences of this discovery unfold in act four (TP 35–37), in which, among other things, Lobachevski finally derives the trigonometric formulae of imaginary geometry that he promised us at the end of TP 22.

With a new sense of its importance, we now turn to Lobachevski's proof of the prism theorem.

Suppose that three planes intersect one another along three parallel lines, AA', BB', and CC' (TP 25). Let X, Y, and Z denote the dihedral angles they form at AA', BB', and CC', respectively. Take random points A, B, and C, one from each line, and construct the plane passing through them. Construct a second plane containing the line AC and some point D of BB'. Let the dihedral angle that this plane makes with the plane containing the parallel lines AA' and CC' be denoted by w.

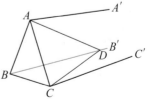

Lobachevski's reference to TP 25 is simply a reminder that parallelism is transitive, so that he does not have to specify which edge is parallel to which. They are all parallel to one another. In terms of the names that Lobachevski gave to the dihedral angles, to prove the prism theorem is to prove that $X + Y + Z = \pi$.

To obtain information about the prism's dihedral angles, Lobachevski introduces two auxiliary constructions. The first, plane $\triangle ACD$, brings three interrelated trihedral angles into play. (Their vertices are at A, C and D.) This will allow him to use TP 27 to extract dihedral information from trihedral sources. In order to get what he needs from these trihedral angles, he introduces his second auxiliary construction: he puts a sphere about each of the trihedral angles' vertices. These will allow him to convert questions about trihedral angles into questions about the spherical triangles that they subtend. Thus, Lobachevski's overall strategy is to deduce information about the prism's dihedral angles by studying certain spherical triangles.

The First Sphere

Draw a sphere centered at A; the points in which the lines AC, AD, and AA' intersect it determine a spherical triangle, whose size we shall denote by α, and whose sides we shall denote p, q, and r. If q and r are those sides whose opposite angles have measures w and X respectively, then the angle opposite side p must have measure $\pi + 2\alpha - w - X$. (TP 27)

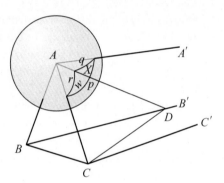

In the passage above, he constructs the first of these spherical triangles; he calls its sides p, q, and r, and he denotes its size by α.

Recall that the angle between two sides of a spherical triangle is equal in measure to the dihedral angle between the planes that contain them (TP 26 Notes, Claim 2). This allows us to find two of the angles of the spherical triangle. The planes containing sides p, q, and r are $AA'\,BB'$, $AA'\,CC'$, and ACD, respectively. Hence, it follows that the angle between p and q is X (since this is the dihedral angle between planes $AA'\,BB'$ and $AA'\,CC'$), and that the angle between p and r is w (since this is the dihedral angle between planes $AA'\,CC'$, and ACD).

To obtain the remaining angle, we can use the reformulation of TP 27 (TP 27 Notes, Claim 1), which tells us that the triangle's *size* equals half of its angular excess. Expressed in symbols, this yields

$$\alpha = \frac{1}{2}[X + w + (\text{the triangle's 3rd angle}) - \pi].$$

Thus, after rearranging this equation, we find that the third angle's measure is $\pi + 2\alpha - w - X$.

Lobachevski now turns to the second and third spheres, centered at C and D.

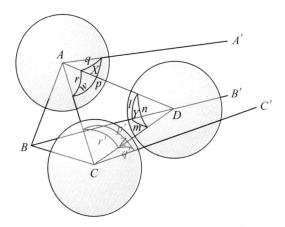

The Other Spheres

"All the other spheres, in varying ways,
direct their distinctive qualities
to their own purposes and influence."
—Dante Alighieri, *Paradiso*, II, 118–120 (tr. Hollander).

Similarly, the intersections of CA, CD, and CC' with a sphere centered at C determine a spherical triangle of size β, whose sides are denoted by p', q', and r', and whose angles are: w opposite q', Z opposite r', and thus, $\pi + 2\beta - w - Z$ opposite p'.

This is the same argument, *mutatis mutandis*, given for the first sphere.

The third sphere, centered at D, requires a slightly different, but essentially similar line of reasoning.

Finally, the intersections of DA, DB, and DC with a sphere centered at D determine a spherical triangle, whose sides, l, m, and n, lie opposite its angles, $w + Z - 2\beta$, $w + X - 2\alpha$, and Y, respectively. Its size, consequently, must be

$$\delta = \frac{1}{2}(X + Y + Z - \pi) - (\alpha + \beta - w).$$

The angle between sides l and m has the same measure as the dihedral angle between planes $AA'BB'$ and $BB'CC'$. Thus, the measure of the angle between l and m is Y.

The angle between sides m and n has the same measure as the dihedral angle between planes BCD and ACD. This dihedral angle, in turn, is the supplement of the dihedral angle that measures the angle between r' and q'. Since this last has measure $\pi + 2\beta - w - Z$, its supplement's measure is $\pi - (\pi + 2\beta - w - Z) = w + Z - 2\beta$. Consequently, the measure of the angle between m and n is $w + Z - 2\beta$.

Similarly, the measure of the angle between l and n is $\pi - (\pi + 2\alpha - w - X) = w + X - 2\alpha$.

Now that we have all three angles, the reformulation of TP 27 (TP 27 Notes, Claim 1) immediately yields the size, δ, of the spherical triangle:

$$\delta = \frac{1}{2}[Y + (w + Z - 2\beta) + (w + X - 2\alpha) - \pi] = \frac{1}{2}(X + Y + Z - \pi) - (\alpha + \beta - w).$$

Rotating the Auxiliary Plane

If w decreases toward zero, then α and β will vanish as well, so that $(\alpha + \beta - w)$ can be made less than any given number. Since sides l and m of triangle δ will also vanish (TP 21), we can, by taking w sufficiently small, place as many copies of δ as we wish, end to end, along the great circle containing m, without completely covering the hemisphere with triangles in the process. Hence, δ vanishes together with w. From this, we conclude that we must have $X + Y + Z = \pi$.

Lobachevski's argument is more obscure here than it ought to be; I shall present what I hope is a clearer version of the same.

We want to prove that $X + Y + Z = \pi$. Since a rearrangement of the terms in the expression for δ that we found above reveals that $X + Y + Z = \pi + 2(\alpha + \beta - w + \delta)$, we can do this by proving that $2(\alpha + \beta - w + \delta) = 0$. Let point D travel down ray BB'. As it moves, the quantities $\alpha, \beta, w,$ and δ will all vary. On the other hand, the quantity $2(\alpha + \beta - w + \delta)$ will not vary, since it equals $X + Y + Z - \pi$, an expression whose value is clearly unaffected by the location of D. The heart of the proof, which we shall examine in a moment, is to show that as D recedes, the quantities $\alpha, \beta, w,$ and δ all approach zero. Once we establish this, the proof will be essentially complete; given any positive number ε, we simply take D far enough away to ensure that $\alpha, \beta, w,$ and δ will all be less than $\varepsilon/8$, thus guaranteeing that

$$|2(\alpha + \beta - w + \delta)| \leq 2(|\alpha| + |\beta| + |w| + |\delta|) \leq 2(4 \cdot \varepsilon/8) = \varepsilon.$$

Since the absolute value of the constant quantity $2(\alpha + \beta - w + \delta)$ is smaller than any positive number ε, the constant $(\alpha + \beta - w + \delta)$ must, in fact, be zero, as claimed. Thus, $X + Y + Z = \pi$, proving the prism theorem.

Let us now attend to the details, and demonstrate that $\alpha, \beta, w,$ and δ all vanish as $D \to \infty$.

Claim 1. As $D \to \infty$, $w \to 0$.

Proof. As D recedes, lines AD and CD rotate about A and C respectively, while plane ACD rotates about AC, causing w to decrease. The limiting position of AD must be AA', the unique parallel to BB' through A. (*i.e.* the first position at which the rotating line fails to cut BB'.) Similarly, CD approaches CC'. Consequently, the rotating plane ACD has plane $AA'CC'$ as its limiting position. Hence, w approaches zero, as claimed. ∎

Claim 2. As $D \to \infty$, α and $\beta \to 0$.

Proof. Consider the first spherical triangle, whose size is α. As $D \to \infty$, two of its vertices remain fixed. These two vertices lie on the great circle in which plane $AA'CC'$ intersects the sphere. Since the third vertex lies on plane ACD, which rotates toward $AA'CC'$, its limiting position as $D \to \infty$ is also a point on this great circle. Thus, in the limit, all three vertices lie on the same great circle, and therefore form a *degenerate* spherical triangle of area zero. Hence, α approaches zero as $D \to \infty$. For similar reasons, β approaches zero as well. ∎

Claim 3. As $D \to \infty$, $\delta \to 0$.

Proof. As D recedes into the distance, the sphere centered about it comes along for the ride; the sphere itself does not change its shape, but the triangle upon it does. Since $\angle ADB$ and $\angle BDC$ both approach zero as $D \to \infty$ (TP 21), the lengths of the arcs they subtend (l and m) also approach zero. Since two sides of the triangle vanish, the third side must vanish with them. Hence, the triangle's area (and hence its size) approaches zero as $D \to \infty$. ∎

Having disposed of these details, we have proved the prism theorem. The picturesque justification for the vanishing of δ in Lobachevski's final sentences amounts to the following: for any natural number N, we can push D sufficiently far away to guarantee that its spherical triangle will be so small that we can paste N non-overlapping copies of it on the sphere's surface without completely covering it. Since N copies won't cover the sphere's surface, δ must be less than $1/N$. Hence, for any positive real number ε, we can choose a whole number N such that $1/N < \varepsilon$ and then push D sufficiently far away to guarantee that $\delta < 1/N < \varepsilon$. In other words, as D goes to ∞, δ goes to zero.

Theory of Parallels 29

In a rectilinear triangle, the three perpendicular bisectors of the sides meet either in a single point, or not at all.

Non-concurrent Perpendicular Bisectors

> "Sit down, the two of you, there before me," said Neary, "and do not despair. Remember there is no triangle, however obtuse, but the circumference of some circle passes through its wretched vertices."
>
> — Samuel Beckett, *Murphy*

In Euclidean geometry, every triangle's perpendicular bisectors are concurrent. In fact, they meet at the triangle's circumcenter (*Elements* IV.5). In contrast, it is easy to construct a triangle in imaginary geometry with non-concurrent perpendicular bisectors.

Suppose that $AB \| CD$. From G, an arbitrary point that lies between the parallels, drop perpendiculars GE and GH, as shown in the figure. Double the lengths of these segments, extending them to F and I, respectively. Notice that F, G, and I cannot be collinear: if they were, then the line upon which they lie would be a common perpendicular for the parallels, which is impossible in imaginary geometry (TP 22). Consequently, these points form the vertices of a triangle, ΔFGI. Since two of the perpendicular bisectors of this triangle are parallel to one another, the three bisectors obviously cannot meet at a point.

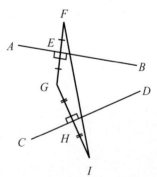

Such triangles may seem like mere curiosities, but they will play major roles in TP 31.

In TP 29, Lobachevski proves that if two of a triangle's perpendicular bisectors meet one another, then all three will be concurrent. The point of concurrence—if it exists—will be the center of a circle that passes through the triangle's vertices.

The Proof

Suppose that two of triangle ABC's perpendicular bisectors, say, those erected at the midpoints E & F of AB and BC respectively, intersect at some point, D, which lies within the triangle. Draw the lines DA, DB, and DC, and observe that the congruence of the triangles ADE and BDE (TP 10) implies that $AD = BD$. For similar reasons, we have $BD = CD$, whence it follows that triangle ADC is isosceles. Consequently, the perpendicular dropped from D to AC must fall upon AC's midpoint, G. This reasoning remains valid when D, the point of intersection of the two perpendiculars ED and FD, lies outside the triangle, or when it lies upon side AC.

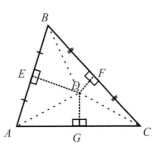

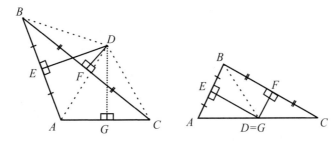

Thus, if two of the three perpendiculars fail to intersect one another, then neither of them will intersect the third.

Notes on the Proof

In the first two cases (when D lies either within or without the triangle), the perpendicular dropped from D to AC splits $\triangle ADC$ into a pair of subtriangles, $\triangle DAG$ and $\triangle DCG$, whose congruence by RASS (TP 10) implies that $AG = CG$. Since G is the midpoint of AC, GD must be the perpendicular bisector of AC. In other words, the perpendicular bisectors ED, FD, and GD all meet at D, as claimed. The third case (when D lies on the side of the triangle) is simpler still: the perpendicular bisectors of AB and BC meet *at* the third side's midpoint.

Finally, it is easy to see that the point of concurrence D (when it exists) is $\triangle ABC$'s *circumcenter*. Since $AD = BD = CD$, as Lobachevski indicates in his proof, the unique circle centered at D that passes through A must also pass through B and C.

Circumcircles, the Parallel Postulate, and Bolyai the Elder

We now add one more item to our list of statements that are equivalent to the parallel postulate.

Claim. The parallel postulate holds if and only if every triangle has a circumcircle.

Proof. $\Rightarrow$) Euclid IV.5.

$\Leftarrow$) Suppose that every triangle has a circumcircle. Let l be a line and $P \notin l$ be a point. Drop the perpendicular PQ to l, and let m be the perpendicular erected upon PQ at P. By Euclid I.28, m does not intersect l. We will now show that all other lines through P must intersect l. Let n be such a line. Let A be an arbitrary point between P and Q. Extend PQ through

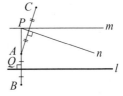

Q to B so that $AQ = QB$. Drop the perpendicular AR to n, and extend it through R to C so that $RC = AR$. Since A, B, and C are noncollinear, they are the vertices of a triangle. Thus, by hypothesis, there exists a circle passing through them. By Euclid III.1 (a neutral theorem), the perpendicular bisectors of any two chords of a circle will meet at the circle's center. Consequently, n (the perpendicular bisector of chord AC) will meet l (the perpendicular bisector of chord AB), which was to be shown. ■

Farkas Bolyai (1775–1856), the father of János, discovered the preceding proof. He devoted much thought to the parallel postulate, but the following excerpts[1] from letters to János suggest that his studies in this area may have had a less than salubrious effect upon his mind.

> You must not attempt this approach to parallels. I know this way to its very end. I have traversed this bottomless night, which extinguished all light and joy of my life. I entreat you, leave the science of parallels alone. . .I thought I would sacrifice myself for the sake of truth. I was ready to become a martyr who would remove the flaw from geometry and return it purified to mankind. I accomplished monstrous, enormous labors; my creations are far better than those of others and yet I have not achieved complete satisfaction . . . I turned back when I saw that no man can reach the bottom of the night. I turned back unconsoled, pitying myself and all mankind. Learn from my example: I wanted to know about parallels, I remain ignorant; this has taken all the flowers of my life and all my time from me.

> You should fear it like a sensual passion; it will deprive you of health, leisure and peace—it will destroy all joy in your life. These gloomy shadows can swallow up a thousand Newtonian towers and never will there be light on earth; never will the unhappy human race reach absolute truth—not even in geometry.

Fortunately, János persisted in his researches, despite his father's wishes.

---

[1] I have taken the first from Gray, *János Bolyai* (p. 51), and the second from Rosenfeld (p. 108).

Theory of Parallels 30

In a rectilinear triangle, if two of the perpendicular bisectors of the sides are parallel, then all three of them will be parallel to one another.

This proposition continues the story of its predecessor. In imaginary geometry, certain triangles lack circumcircles, since their perpendicular bisectors fail to meet. In TP 29, we saw that if two of the perpendicular bisectors of a triangle's sides intersect one another, then all three bisectors must be concurrent. But what happens if no two bisectors meet? In TP 30, Lobachevski gives a partial answer: if two bisectors are not only non-intersecting, but also *parallel* to one another, then the third bisector will be parallel to them as well.

His proof falls into two cases: one in which the two given parallels lie on opposite sides of the third perpendicular bisector, and one in which they lie on the same side of it.

The First Case

In triangle $\triangle ABC$, erect perpendiculars DE, FG, and HK from D, F, and H, the midpoints of the sides. (See the figure.)

We first consider the case in which DE and FG are parallel, and the third perpendicular, HK, lies between them. Let L and M be the points in which the parallels DE and FG cut the line AB. Draw an arbitrary line entering angle $\angle BLE$ through L. Regardless of how small an angle it makes with LE, this line must cut FG (TP 16); let G be the point of intersection. The perpendicular HK enters triangle

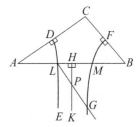

$\triangle LGM$, but because it cannot intersect MG (TP 29), it must exit through LG at some point P. From this it follows that HK must be parallel to DE (TP 16 & 18) and FG (TP 18 & 25).

In this first case, we suppose that two of the perpendicular bisectors are parallel ($DE\|FG$) and lie on opposite sides of the remaining perpendicular bisector (HK). Lobachevski's straightforward proof that $HK\|DE$ and $HK\|FG$ makes use of parallelism's symmetry and transitivity (TP 16, 18, 25): he shows that $DE\|HK$, at which point symmetry gives $HK\|DE$, whence transitivity yields $HK\|FG$.

The second case proves surprisingly stubborn. Before proving it, Lobachevski pauses to record a few formulae, whose purpose will become clear shortly.

Lobachevski's Observations: Three Formulae

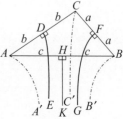

In the case just considered, if we let the sides $BC = 2a$, $AC = 2b$, $AB = 2c$, and denote the angles opposite them by A, B, C, we can easily show that

$$A = \Pi(b) - \Pi(c) \quad B = \Pi(a) - \Pi(c) \quad C = \Pi(a) + \Pi(b)$$

by drawing lines AA', BB', CC', from points A, B, C, parallel to HK—and therefore parallel to DE and FG as well (TP 23 & 25).

"The case just considered" refers to the circumstance in which all three perpendicular bisectors are parallel to one another. In this scenario, one can verify the three formulae simply by looking at the figure. For instance, $B = \angle FBH = \angle FBB' - \angle HBB' = \Pi(a) - \Pi(c)$.

In fact, the validity of each individual formula is equivalent to the parallelism of a particular pair of bisectors. For example, our derivation of the formula $B = \Pi(a) - \Pi(c)$ depends only upon the fact that $HK \| FG$. Conversely, if $B = \Pi(a) - \Pi(c)$ is known to hold, we can prove that $HK \| FG$.[1]

In this manner, one can show that:

$$A = \Pi(b) - \Pi(c) \quad \Leftrightarrow \quad HK \| DE.$$
$$B = \Pi(a) - \Pi(c) \quad \Leftrightarrow \quad HK \| FG.$$
$$C = \Pi(a) + \Pi(b) \quad \Leftrightarrow \quad DE \| FG.$$

The Second Case

Next, consider the case in which HK and FG are parallel. Since DE cannot cut the other two perpendiculars (TP 29), it either is parallel to them, or intersects AA'.

In the second case, we suppose that two of the bisectors are parallel ($HK \| FG$) and lie on the same side of the remaining perpendicular bisector (DE). We must show that the remaining bisector is parallel to the first two. To do so, it will suffice to show that $DE \| HK$: for if we can establish this, then the transitivity of parallelism will imply that $FG \| HK$. Lobachevski proves that $DE \| HK$ by a *reductio ad absurdum* argument.

First, DE cannot cut HK: the intersection of two bisectors would force all three bisectors to be concurrent (TP 29). In particular, HK and FG would intersect one another, contradicting the fact that they are parallels.

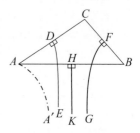

Thus, DE must be either parallel or ultraparallel to HK.

If DE is ultraparallel to HK, it must, according to Lobachevski, intersect AA' (which is defined as the line drawn through A parallel to HK). Lobachevski offers no proof of this fact, presumably because

---

[1] *Proof.* $BB' \| HK$ by definition, so $\angle B'BH = \Pi(c)$. Thus, we can rewrite $B = [\Pi(a) - \Pi(c)]$ as $B = [\Pi(a) - \angle B'BH]$. Solving for $\Pi(a)$, we obtain $\Pi(a) = [B + \angle B'BH] = \angle B'BF$, which implies that $BB' \| FG$. Thus, $HK \| FG$ by transitivity.

he felt that the proof was obvious. What "obvious" proof did Lobachevski see in his mind's eye? The most plausible candidate that I can think of (that uses only ideas we have developed thus far) is the following.

Claim 1. If $FG \, \| HK$, but DE is not parallel to HK, then DE intersects line AA'.

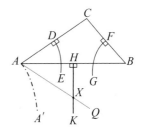

Proof. DE and AA' are not parallel. For, if they were, then DE would be parallel to HK, by transitivity, contrary to hypothesis.

DE and AA' are not ultraparallel. If they were, then we could rotate AA' slightly about A to obtain a line AQ that enters $\angle A'AH$, but does not cut DE. Because AQ enters $\angle A'AH$, it would cut HK at some point X, since $AA' \| HK$. Since DE enters $\triangle AXH$ through side AH, it would have to exit through one of the triangle's remaining sides, which, however, is impossible: by the very definition of AQ, line DE cannot intersect $AQ = AX$; moreover, by TP 29, if DE were to cut its fellow perpendicular bisector $HX = HK$, the parallels FG and HK would meet, which is absurd.

Since DE and AA' are neither parallel nor ultraparallel, they must intersect one another, as claimed. ■

To be fussy, even *this* argument rests upon the assumption that DE must cross segment AH. Plausible though this sounds, it is always a little bit dangerous to dismiss anything as "obvious" in the counterintuitive world of imaginary geometry. But fear not: this little hole *can* be safely filled. Filling it, however, would entail a page-long digression, so I leave this as an exercise for faithless readers, lest the rest of us lose the thread of Lobachevski's argument.

The Second Case Continued

Recall that we are trying to prove that DE is parallel to HK. Since DE cannot cut HK (by TP 29), we know that DE is either parallel or ultraparallel to HK. Thus, to prove that the lines are parallel, it suffices to show that they cannot be ultraparallel. As a first step in this direction, we demonstrated that if DE were ultraparallel to HK, then DE would have to intersect AA' (the line through A drawn parallel to HK and FG).

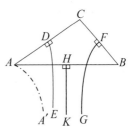

Consequently, it remains to show that DE *cannot* intersect AA'. This will prove that DE cannot be ultraparallel to HK, and hence, that DE must be parallel to HK, which is what we want to demonstrate.

Lobachevski's first step is to secure an inequality.

To assume this latter possibility is to assume that $C > \Pi(a) + \Pi(b)$.

That is, to assume that DE intersects AA' (the hypothesis Lobachevski is trying to destroy) would imply that $C > \Pi(a) + \Pi(b)$. We can prove this as follows.

Claim 2. If DE and AA' meet, then $C > \Pi(a) + \Pi(b)$.

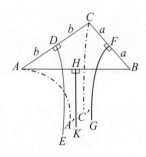

Proof. Recall that CC' is parallel to HK and FG, by definition.
From the figure, we see that $C = (\angle FCC' + \angle C'CD) = (\Pi(a) + \angle C'CD)$.

Thus, to prove the claim amounts to proving that $\angle C'CD > \Pi(b)$.

If $\angle C'CD = \Pi(b)$, then $DE \| CC' \| AA'$, which is absurd: inter-
secting lines cannot be parallel. If $\angle C'CD < \Pi(b)$, then CC' cuts
DE. Thus, ray DE intersects both AA' and CC', which is impossi-
ble: it emanates from a point lying between these two parallel lines, and thus may cut only one
of them. (To meet the second line, it would have to recross the first, in violation of TP 2.) Thus,
we must have $\angle C'CD > \Pi(b)$. ∎

If this is the case, we can decrease the magnitude of this
angle to $\Pi(a) + \Pi(b)$ by rotating line AC to a new position
CQ (see the figure). The angle at B is thereby increased.
That is, in terms of the formula proved above,

$$\Pi(a) - \Pi(c') > \Pi(a) - \Pi(c),$$

where $2\,c'$ is the length of BQ. From this it follows that $c' > c$ (TP 23).

Rotating AC about C, Lobachevski transforms $\triangle ABC$ into $\triangle QBC$, a new triangle with a
strategically constructed angle: $\angle BCQ = \Pi(a) + \Pi(b)$. Since Q and C lie on opposite sides of
AB, we have $\angle QBC > \angle ABC$. That is, the angle at B in the new triangle is larger than it was
in the old triangle. In his quest for a contradiction, Lobachevski expresses the two angles at B
in terms of the side lengths. Here, he will finally use the formulae that he described between the
two cases of his proof.

Observation. $\angle ABC = \Pi(a) - \Pi(c)$.

Proof. As noted above (in "Lobachevski's Observations"), this formula follows from the
fact that $HK \| FG$. ∎

Claim 3. $\angle QBC = \Pi(a) - \Pi(c')$.

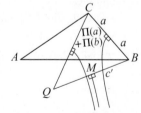

Proof. Since angle C in $\triangle QBC$ equals $\Pi(a) + \Pi(b)$, the perpendicular bisectors of the sides adjacent to this angle must be parallel
to one another, as noted above. Both of these parallel bisectors must
leave the triangle through side QB. If M is the midpoint of QB, we
can prove that one of the bisectors must leave through QM, and the
other through MB as follows.

Suppose, by way of contradiction, that both leave through the
same half of QB (say MB). Then we would have $MQ < MC$ (since M would lie on the A-side
of the plane as partitioned by the perpendicular bisector of QC). Similarly, $MC < MB$. Thus,
we have $MQ < MB$, which contradicts the fact that M is the midpoint of QB.

Hence, *M* lies *between* the two parallel bisectors. Consequently, by the first case of the proof, the perpendicular bisector of QB must be parallel to the other two bisectors. Since all three perpendicular bisectors of $\triangle QBC$ are parallel to one another, all three of Lobachevski's formulae hold in this triangle. In particular, the angle at *B* will be $\Pi(a) - \Pi(c')$, as claimed. ∎

Since $\angle QBC > \angle ABC$, the expressions from the preceding Observation and Claim tell us that

$$\Pi(a) - \Pi(c') > \Pi(a) - \Pi(c),$$

which implies that $\Pi(c') < \Pi(c)$. Since Π is a decreasing function (TP 23 Notes, Claim 1), it follows that $c' > c$. And yet...

On the other hand, since the angles at *A* and *Q* in triangle $\triangle ACQ$ are equal, the angle at *Q* in triangle $\triangle ABQ$ must be greater than the angle at *A* in the same triangle. Consequently, $AB > BQ$ (TP 9); that is, $c > c'$.

Because $\triangle ACQ$ is isosceles, the base angles $\angle CAQ$ and $\angle CQA$ are equal (Euclid I.5). Let θ be their common measure. In $\triangle ABQ$, it is clear that $\angle BAQ < \theta < \angle AQB$. Thus, $QB < AB$ (Euclid I.19). That is, $2c' < 2c$, or $c > c'$. We have arrived at a contradiction, having demonstrated that *c* is simultaneously less than and greater than c'.

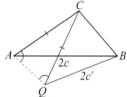

This contradiction followed from the assumption that DE intersects AA'. Early in the second case, Lobachevski proved that DE either intersects AA' or is parallel to the other perpendicular bisectors. Having disposed of the former possibility, we conclude that the latter must be true. This concludes the second case, and with it, at last, the proof of TP 30.

Points at Infinity

By adopting the convention that parallel lines meet at a "point at infinity" in their direction of parallelism, we can unify the statements of Lobachevski's 29th and 30th propositions as follows: *if two perpendicular bisectors of a rectilinear triangle meet at a point (possibly at infinity), then the third must pass through the same point as well.*

If a triangle's perpendicular bisectors meet at a point at infinity, might the triangle have a circumcircle in some extended sense? A circle whose center is at infinity? What can we say about the nature of such a figure, if anything at all? In the early 17th-century, Johannes Kepler professed that a circle whose center is at infinity is a straight line. Our hypothetical infinite circumcircle, however, cannot be straight, for the simple reason that no straight line can pass through all three of a triangle's vertices.

In the next two propositions, we shall make the acquaintance of the curve that plays the role of circumcircle in this situation: the *horocycle*, a curve neither straight nor circular, but enjoying both line-like properties and circle-like properties.

Theory of Parallels 31

We define a horocycle to be a plane curve with the property that the perpendicular bisectors of its chords are all parallel to one another.

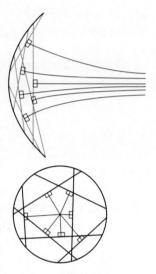

Horocycles

It is often convenient to think of a horocycle as a "circle of infinite radius", or "a circle whose center is at infinity", but this will not suffice as a formal definition. Although one typically defines a circle as the locus of points at a fixed distance from a given point, Lobachevski used an alternate, equivalent definition of a circle as the basis for his definition of the horocycle. Namely, a circle is "a closed curve in the plane with the property that the perpendicular bisectors of its chords are all concurrent," and its center is the point of concurrence. (Circles clearly possess this property, and one may prove that any closed plane curve exhibiting it is circular.) Since a horocycle's center is supposed to be "at infinity", it ought to be a curve whose chords' perpendicular bisectors meet there. Worded more rigorously, it ought to be a plane curve whose chords' perpendicular bisectors are all *parallel*. This is precisely how Lobachevski defines it.

One fussy detail remains. The topological clause in the alternate definition of a circle (a circle must be *closed*) is designed to prevent mere circular *arcs* from satisfying the definition. For horocycles, the filter of closure is too fine; it would keep out not only horocyclic arcs, but full horocycles as well, since even they are not closed curves, as we shall see shortly. Lobachevski, in a minor oversight, fails to supply an appropriate filter, but we can easily remedy this by declaring that *no horocycle can be a proper subset of another.*

Generating a Horocycle

In accordance with this definition, we may imagine generating a horocycle as follows: from a point A on a given line AB, draw various chords AC of length $2a$, where $\Pi(a) = \angle CAB$. The endpoints of such chords will lie on the horocycle, whose points we may thus determine one by one.

We can define a unicorn, but this does not imply that such creatures actually exist. Having defined horocycles, Lobachevski hastens to exhibit one. His process for generating a horocycle is, I believe, easier to understand in the following dynamic reinterpretation.

92

Given a ray AB, which we shall call the *axis* of the horocycle, erect a perpendicular ray from A, and let it slowly rotate toward AB, so that the angle that it makes with AB decreases from $\pi/2$ to 0. Let C be a moving point, initially coincident with A, which moves down the rotating ray as it turns. The following rule governs the motion of C: if θ is the angle that the rotating ray makes with AB, and a is the length such that $\Pi(a) = \theta$ (such a length exists for every θ by TP 23), then C will be at a distance of $2a$ from A. Thus, $AC \to \infty$ as $\theta \to 0$ (TP23). The moving point C will trace out half of a horocycle; we shall *prove* that it actually satisfies the definition in a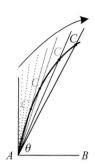

moment. The horocycle's other half, the mirror image of the first, can be obtained by carrying out the same procedure, but beginning with the *other* ray emanating from A and perpendicular to AB.

Next, we shall verify that the curve traced out by C satisfies the horocycle's definition.

The perpendicular bisector DE of a chord AC will be parallel to the line AB, which we shall call the *axis* of the horocycle. Since the perpendicular bisector FG of any chord AH will be parallel to AB, the perpendicular bisector KL of any chord CH will be parallel to AB as well, regardless of the points C and H on the horocycle between which the chord is drawn (TP 30). For that rea-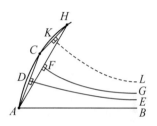

son, we shall not distinguish AB alone, but shall instead call *all* such perpendiculars *axes of the horocycle*.

These, Lobachevski's concluding words for TP 31, fail to do justice to the properties of the curve that he has just brought to light. In these notes, we shall explore its properties in greater depth than Lobachevski does, so that we might obtain more insight into its nature. But first, let us carefully verify that the curve he has generated is indeed a horocycle.

Claim 1. The curve traced out by C is indeed a horocycle.

Proof. We must show that the perpendicular bisectors of all the curve's chords are parallel to one another. Thanks to the transitivity of parallelism, it suffices to show that they are all parallel to AB. By the curve's construction, it is clear that the perpendicular bisector of any chord AC that joins A to any other point on the curve will be parallel to AB. Any chord CH that joins two points on the curve, neither of which is A, should be thought of as one side of the triangle $\triangle ACH$; since the perpendicular bisectors of the other two sides (AC and AH) are parallel to AB, the perpendicular bisector of CH must also be parallel to it, by TP 30. Hence, the curve traced out by C satisfies Lobachevski's definition of a horocycle, and thus lies within some complete horocycle $\mathcal{H}$. If we can prove that $\mathcal{H}$ lies within the curve traced out by C, we will be able to conclude that the curve traced out by C is identical to $\mathcal{H}$. We do this now.

The perpendicular bisectors of chords of $\mathcal{H}$ are, by the definition of a horocycle, parallel to one another. Since some of them are known to be parallel to AB (namely, those bisecting chords that join points on the curve traced out by C), the transitivity of parallelism implies that *all* bisectors of $\mathcal{H}$'s chords are parallel to AB. In particular, if P is an arbitrary point on $\mathcal{H}$, then MM', the perpendicular bisector of AP, will be parallel to AB, making $\angle BAM = \Pi(AM)$. It follows that P lies on our curve: when the rotating ray makes the angle $\Pi(AM)$ with axis AB during the curve's generation, the curve acquires P, since it is the point on the ray at a distance of $2(AM)$ from A. Hence, every point of the horocycle $\mathcal{H}$ lies within the curve traced out by C. Having shown that the two curves are identical, we conclude that the curve traced out by C is a horocycle, as claimed.   ∎

Thus, horocycles exist; we may generate one at will simply by choosing a ray as an axis and following the procedure described above.

Euclid's construction of an equilateral triangle (Euclid I.1) proves that triangles exist, but it is obviously insufficient for constructing *all* triangles; equilaterals constitute but one species of the larger triangle genus. In contrast, we can prove that Lobachevski's generation of a horocycle is comprehensive: we can obtain *every* horocycle this way. That is, we can demonstrate that every horocycle $\mathcal{H}$ has an axis AB – a ray that yields $\mathcal{H}$ when we carry out Lobachevski's generation process upon it.

Every Horocycle has an Axis

Claim 2. Every horocycle has an axis.

Proof. To be more specific, we shall show that any ray that emanates from a point on a horocycle and is parallel to all perpendicular bisectors of the horocycle's chords is necessarily an axis of that horocycle.

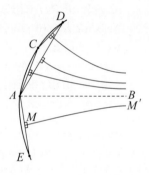

To this end, let $\mathcal{H}$ be a horocycle, and let A, C, and D be arbitrary points upon it. By definition of a horocycle, the perpendicular bisectors of AC, AD, and CD (and all other chords, for that matter) are parallel to one another. Let AB be the ray that emanates from A and is parallel to these bisectors. In fact, the perpendicular bisector of *every* chord of $\mathcal{H}$ will be parallel to AB, by definition of a horocycle, together with the transitivity of parallelism.

Let $\mathcal{K}$ be the horocycle whose axis is AB. We shall show that $\mathcal{H} = \mathcal{K}$.

Let E be an arbitrary point of $\mathcal{H}$. Since $MM' \| AB$ (where MM' is the perpendicular bisector of AE), we have $\angle BAM = \Pi(AM)$. Now, when the rotating ray that generates $\mathcal{K}$ makes angle $\Pi(AM)$ with AB, $\mathcal{K}$ acquires the point on that ray which lies at distance $2(AM)$ from A; that is, $\mathcal{K}$ acquires point E. Hence, E lies on $\mathcal{K}$, so $\mathcal{H} \subseteq \mathcal{K}$. By definition, one horocycle cannot be a proper subset of another, so $\mathcal{H} = \mathcal{K}$. Thus, since AB is an axis for $\mathcal{K}$, AB is also an axis for $\mathcal{H}$.   ∎

Having proved the non-self-evident truth that all horocycles are created equal, we turn to the properties with which they are endowed. The proposition just established immediately yields two that are very striking.

Many Axes, Much Symmetry

Claim 3. Every horocycle has infinitely many axes—one through each of its points.

Proof. In the preceding claim, point A was entirely arbitrary. Since the axis that we constructed emanated from A, we could just as easily have constructed an axis emanating from any other point of $\mathcal{H}$.   ∎

Claim 4. Horocycles possess a tremendous amount of symmetry: a horocycle is symmetric about each of its axes.

Proof. Clearly, the process of generating a horocycle yields a curve that is symmetric about its generating axis. Every horocycle has an axis (Claim 2), and thus every horocycle has a line of reflective symmetry. Since Claim 3 tells us that horocycles have infinitely many axes, they also have infinitely many lines of symmetry. ■

If we think of a horocycle as a circle of infinite radius, then its axes play the role of diameters. Thus, for example, Claim 4 is analogous to the fact that circles are symmetric about all of their diameters. Of course, horocycles are not entirely circle-like. Circles come in a variety of sizes, but the same is not true of horocycles, as we now show.

If You've Seen One, You've Seen Them All

Claim 5. All horocycles are congruent.

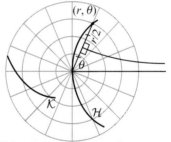

Proof. Imagine the plane as a piece of paper upon which two horocycles, $\mathcal{H}$ and $\mathcal{K}$, are drawn. Introduce polar coordinates by drawing a polar grid on an overhead transparency and laying it upon the plane, making its polar axis (the ray $\theta = 0$) coincide with an axis of $\mathcal{H}$. Lobachevski's method for generating horocycles endows $\mathcal{H}$ with the polar equation $\theta = \Pi(r/2)$. Of course, by repositioning the transparency, we can supply $\mathcal{K}$ with this equation just as easily. Since $\mathcal{H}$ and $\mathcal{K}$ have the same equation up to an isometric (distance preserving) change of coordinates, they must be congruent. (Our change of coordinates is isometric since it merely involves laying down the polar grid, the object that defines the distances in our equation, in two distinct places.) ■

We are used to such uniformity among points and lines, the basic elements of geometry, but to find this sameness among horocycles is quite remarkable.

The "thought experiment" with the polar grid has a second notable consequence. If, instead of moving the polar axis from an axis of $\mathcal{H}$ to an axis of $\mathcal{K}$, we move it to a second axis of $\mathcal{H}$, then the polar equation of $\mathcal{H}$ will remain the same (although the coordinates of its individual points will obviously change). To appreciate the significance of this fact, first note that we may think of a curve's equation as a map of its features, where each solution of the equation tells us the precise location of one of the curve's points. For example, if its equation has the solution (20, 30), we may interpret this as the instruction: "Go to the origin, look down the polar axis, turn 30° counterclockwise, and walk forward 20 units. You'll find a point of the curve there." Of course, one must know where the origin and polar axis are to use these directions. However, since a horocycle's equation remains the same no matter which of its points we chose as the origin, we do *not* have to know where the origin is to use such directions; once we know the equation, any starting point will do. Thus, horocycles are not only congruent to one another, but each individual horocycle is homogeneous: it "looks the same" from the perspective of any point upon it.

Further Attributes of Horocycles

Because horocycles have polar equations, $\theta = \Pi(r/2)$, which are governed by the continuous function Π (see notes to TP 23), we know that *horocycles are continuous curves*.

We have already shown that imaginary geometry admits trios of points that are neither collinear nor concyclic: the vertices of any triangle whose perpendicular bisectors do not meet. We may harvest such trios with ease from a horocycle; take any three of its points, and you will have one.

Claim 6. If A, B, and C are distinct points on a horocycle, they are not collinear.

Proof. If they were collinear, then the line upon which they lie would be a common perpendicular to two parallel lines (the perpendicular bisectors of AB and BC, which are parallel to one another by the definition of a horocycle), contradicting TP 22 (parallels never share a common perpendicular in imaginary geometry.) ∎

Claim 7. If A, B, and C are distinct points on a horocycle, they are not concyclic.

Proof. A circle passing through all three of them would be the circumcircle of triangle $\triangle ABC$, whose center would necessarily be the point at which the triangle's perpendicular bisectors meet. By definition of the horocycle, there is no such point, and thus, no such circle. ∎

All axes of a horocycle are parallel to one another, but we shall now prove something stronger: every line parallel to an axis of a horocycle is an axis itself. (Properly speaking, every such line *contains* an axis, since we defined an axis to be a ray. However, we shall often abuse the terminology by using the word "axis" to refer to the line containing the ray. The context will always make it clear whether the ray or the line is meant.)

Lemma. *If $\mathcal{H}$ is a horocycle with an axis AB, then every line parallel to AB intersects $\mathcal{H}$.*

Proof. Suppose there is a line MN, parallel to AB, which does *not* intersect $\mathcal{H}$. Because MN and AB draw closer to one another in their direction of parallelism (TP 24), there is a point E on AB (perhaps light-years from A) that is closer to line MN than it is to point A. That is, if EG is the perpendicular dropped from E to MN, then we will have $EA > EG$.

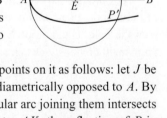

Consider the circle with center E and radius EA. Define two points on it as follows: let J be the point where ray EG intersects it, and let K be the point on it diametrically opposed to A. By construction, J and K lie on opposite sides of $\mathcal{H}$. Hence, the circular arc joining them intersects $\mathcal{H}$ at some point P. Since the circle is symmetric about its diameter AK, the reflection of P in AK lies on it. Moreover, since $\mathcal{H}$ is symmetric about its axis AK, the reflection of P in AK lies upon $\mathcal{H}$ as well. Thus, we have three points, A, P, and P', the last being the reflected image of P in AK, at which $\mathcal{H}$ and the circle meet. This, however, is impossible, since no three points of a horocycle can be concyclic.

Thus, $\mathcal{H}$ intersects every line parallel to AB, as claimed. ∎

Claim 8. If $\mathcal{H}$ is a horocycle with an axis AB, then every line parallel to AB is an axis of $\mathcal{H}$.

Proof. We have just seen that every line parallel to AB cuts $\mathcal{H}$. Thus, any such line contains a ray that emanates from a point on $\mathcal{H}$ and is parallel to AB (and hence is parallel to all the perpendicular bisectors of $\mathcal{H}$'s chords). We have already showed that a ray with these characteristics must be an axis of $\mathcal{H}$. (See the proof of Claim 2 above)   ∎

Corollary. *If $\mathcal{H}$ is a horocycle with an axis AB, then the complete set of its axes is the family of all lines parallel to AB.*

As a result of this corollary, every horocycle is associated with a particular *pencil of parallels*, another name for the set of all lines parallel to a given line in a given direction. We may formalize the notion of "points at infinity" by declaring that two lines "meet at a point at infinity" precisely when they belong to the same pencil of parallels. With this understanding, we may say that each horocycle is associated, via its axes, with a particular point at infinity. We shall call this point the *center* of the horocycle. Thus, intuitively, the center of a horocycle is a point at infinity where its axes meet, while formally, the center is not a point at all, but rather the set of the axes themselves. This convention allows us to prove the following analog of Euclid's third postulate.

Claim 9. Given an ordinary point A and a point at infinity, there exists a horocycle passing through A, whose center is the given point at infinity.

Proof. We must show that there exists a horocycle through A, whose axes are the lines of a particular pencil of parallels. Among these parallels, a unique one passes through A. Let $\mathcal{H}$ be the horocycle generated by it. By Claim 8, its axes must be the lines of the pencil. Thus, $\mathcal{H}$ is a horocycle through A whose center is the given point at infinity.   ∎

Tangents to Horocycles

Although it is difficult to define a *tangent line* to an arbitrary curve without recourse to the language of calculus, certain specific curves possess unmistakable "natural" tangents. For example, the tangent to a circle at any of its points is the unique line passing through it that does not cut the circle a second time. Of course, this line satisfies the calculus definition of tangency as well, but one should not suppose that such "natural tangents" to circles and other conics, which have been known for thousands of years, were somehow illegitimate until they were formally sanctioned by calculus. On the contrary, these tangents lend the calculus definition some of its own authority; if the calculus definition did not agree with the classical definitions in the special cases for which tangents were already known, mathematicians never would have accepted it.

Since horocycles are related to circles, it is not surprising that they too possess natural tangents, which we can identify without calculus. Given any point A on a horocycle, we will show that there is a unique line passing through it that satisfies the following property: every point of the horocycle (other than A) lies on one side of it. Naturally, we will define this distinguished line to be the horocycle's tangent line at A. Just as a circle's tangents are perpendicular to its radii, a horocycle's tangents are perpendicular to its axes, as we shall now demonstrate.

Claim 10. If AB is an axis of a horocycle, then the line erected perpendicular to it at A is the tangent line to the horocycle at A. (That is, it is the unique line through A such that every other point of the horocycle lies to one side of it.)

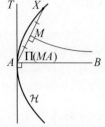

Proof. Let AB be an axis of a horocycle, $\mathcal{H}$. Draw $AT \perp AB$.

First, we shall show that every point of $\mathcal{H}$ lies to one side of line AT.

Let $X \in \mathcal{H}$, and let M be the midpoint of AX. The perpendicular bisector of AX is parallel to AB, so $\angle XAB = \angle MAB = \Pi(MA) < \pi/2$, since the angle of parallelism of any length is acute. Because the side of AT in which B lies consists of the set of points P making $\angle PAB < \pi/2$, the fact $\angle XAB < \pi/2$ implies that X lies on this side on AT. Since X was an arbitrarily chosen point of $\mathcal{H}$, the entire horocycle $\mathcal{H}$ lies on this side of AT, with the exception of A, where the horocycle touches the line.

It remains to show that no other line through A has this property.

Obviously, AB does not satisfy the property.

Suppose that l is any other line through A. (i.e., l is neither AT nor AB.) Any such line must contain points on both sides of line AT. Assume, without loss of generality, that it contains a point Y in the interior of $\angle BAT$. Since $\angle BAY$ is an acute angle, there is a length p such that $\Pi(p) = \angle BAY$ (TP 23). Let Z be the point on ray AY such that $AZ = 2p$. Lobachevski's method for generating the unique horocycle with axis AB guarantees that point Z lies on this horocycle. That is, $Z \in \mathcal{H}$. It follows that l cannot satisfy the required property, for it contains two distinct points of $\mathcal{H}$: A and Z.

Having shown that AT is the only line that satisfies the property, we are justified in calling it the tangent line.   ■

Corollary. *A horocycle cuts its axes perpendicularly.*

Proof. The angle between an axis of a horocycle and the horocycle itself at a given point is defined to be the angle between the axis and the tangent to the horocycle at their point of intersection. By Claim 10, the axis and the tangent are perpendicular to one another. Thus, the axis and horocycle meet at right angles, as claimed.   ■

Euclidean Horocycles?

Lobachevski modeled his definition of a horocycle on a defining property of *circles*, but if we seek the curves that satisfy this definition in Euclidean geometry, we will find ordinary *straight lines*. That is, in the presence of the parallel postulate, horocycles and straight lines are one and the same.

Claim 11. In Euclidean geometry, horocycles are straight lines.

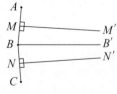

Proof. In Euclidean geometry, lines clearly satisfy the horocycle definition. Conversely, given a curve that satisfies it, let, A, B, and C be three points on it. Let MM' and NN' be the respective perpendicular bisectors of AB and BC. Draw BB' parallel to them. We are working in Euclidean geometry, so we may invoke Euclid I.29 (the first proposition in the *Elements* that depends upon the parallel postulate): when a transversal cuts a pair of parallels, the interior angles on each side of the transversal sum to π. Applying this to the parallels MM' and BB' (with transversal MB), and again to the parallels BB' and NN' (with transversal BN), we find that

$$\angle ABC = \angle MBB' + \angle NBB' = \pi/2 + \pi/2 = \pi.$$

Thus, A, B, and C are collinear. Since the three points were arbitrary, the horocycle must be a straight line, as claimed. ∎

We shall summarize the line-like and circle-like properties of horocycles in the notes to TP 32. Before turning to this proposition, we shall consider an equivalent definition of the horocycle given by Bolyai, and a related idea of Gauss.

Bolyai's L-Curve

> "Bolyai is a much neater workman than Lobachevski, but his work is a little repellent at first owing to the adoption of a strange symbolism of his own invention."
>
> —J.L. Coolidge[1]

Coolidge's assessment on the relative "neatness" of the work of Lobachevski and Bolyai is debatable, but the latter is undeniably a *faster* workman than the former. In §11 of his *Appendix*, four pages after his definition of parallelism, Bolyai introduces a curve named L, and a surface with the equally descriptive name F. These, in Lobachevski's terminology, turn out to be the horocycle and horosphere, respectively. (We shall meet the horosphere in TP 34.) To explain Bolyai's definition of the L-curve, we must first explain some of his "strange symbolism".

According to Bolyai, "$AB \backsimeq CD$ denotes $\angle CAB = \angle ACD$." Unfolding this definition, we see that Bolyai writes $AB \backsimeq CD$ when the rays AB and CD are equally inclined toward the line segment AC. When the rays are equally inclined toward AC *and* parallel to one another, Bolyai writes $AB||| \backsimeq CD$.

Bolyai defines an L-curve by listing its points: given a generating ray AM, which he, like Lobachevski, calls an *axis*, Bolyai declares that A is a point on L, as is the endpoint B of any ray BN such that $BN ||| \backsimeq AM$.

Claim 12. Bolyai's L-curves are horocycles.

Proof. Let L be an L-curve, with axis AM. We shall show that the perpendicular bisectors of its chords are all parallel to AM. This will require two cases.

Case 1. Consider a chord AB, joining A to an arbitrary point B on L. Let PQ be its perpendicular bisector. If BN is the ray such that $BN||| \backsimeq AM$, the equal angles $\angle NBA$ and $\angle MAB$ clearly must be acute for BN and AM to be parallel. To prove that $PQ||AM$, we shall demonstrate that the alternatives lead to contradictions.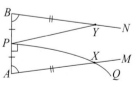

PQ *cannot cut* AM: if it did, say at X, then let $Y \in BN$ be such that $BY = AX$. Then $\triangle PAX \cong \triangle PBY$ by SAS, so $\angle BPY = \angle APX = \pi/2$. That is, PY and PX are both

---

[1] Coolidge, p. 72.

perpendicular to BA. Since there is only one line perpendicular to a given line at a given point, we must have $PY = PX$, which is absurd. Thus, PQ cannot cut AM. By symmetry, PQ cannot cut BN either.

PQ cannot be ultraparallel to AM: if it was, then we could rotate AM slightly toward PQ without causing these initially non-intersecting lines to intersect. Let us try. If we rotate AM toward PQ (and hence toward BN), AM will cut BN at some point X, since $AM||BN$. Ray PQ enters $\triangle ABX$ through side AB, and therefore 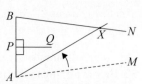 exits through another side (Pasch's axiom). We have already shown that PQ cannot cut $BN(=BX)$, so PQ must exit through AX, the rotated line. Hence, any rotation of AM toward PQ forces an intersection of these lines, so they cannot be ultraparallel.

Therefore, $PQ||AM$, as claimed.

Case 2. Consider a chord BC of L, neither endpoint of which is A. If A, B, and C are noncollinear, we can form the triangle $\triangle ABC$. Since the perpendicular bisectors of the sides AB and AC are parallel to AM (case 1), TP 30 implies that the perpendicular bisector of BC must also be parallel to AM, as was to be shown.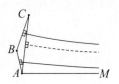

There is no need to consider a second sub-case in which A, B, and C are collinear, for such a configuration cannot occur, as we shall now demonstrate.

Suppose, by way of contradiction, that A, B, and C are collinear (see the figure). Let BN and CO be the rays such that 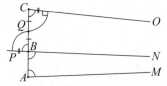 $BN|||\simeq AM$ and $CO|||\simeq AM$. Then $\angle NBA = \angle MAC = \angle OCA$. Bisect BC at Q, and drop a perpendicular QR to CO. Extend ray BN backwards through B to P so that $BP = CR$. Since $\triangle CQR \cong \triangle BQP$ (by SAS), we have $\angle CQR = \angle BQP$. It follows that

$$\angle RQP = \angle RQB + \angle BQP = (\pi - \angle CQR) + \angle BQP = (\pi - \angle BQP) + \angle BQP = \pi.$$

Hence, R, P, and Q are collinear. Since $\angle BPQ = \angle CRQ = \pi/2$, line PR is a common perpendicular for CO and PN, which is impossible since these lines are parallel (TP 22).

Having shown that the perpendicular bisectors of all chords of an L-curve are parallel to one another, we conclude that every L-curve is contained within a horocycle. It remains only to show that every L-curve is a complete horocycle. To this end, let L be an L-curve with axis AM, and let $\mathcal{H}$ be a horocycle in which is contained. We must show that every 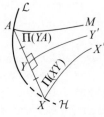 point of $\mathcal{H}$ is also a point of L. Let $X \in \mathcal{H}$, and let XX' be the axis of $\mathcal{H}$ through X. To prove that $X \in L$, we must demonstrate that $XX'|||\simeq AM$.

These lines are clearly parallel, since they are both axes of the horocycle (in our two cases above, we proved that the axis of an L-curve is an axis of the horocycle within which the L-curve lies), so it remains only to show that they are equally inclined toward the chord AX. The perpendicular bisector YY' of this chord is, by definition of a horocycle, parallel to the rest of the horocycle's axes. Thus, $AM||YY'||XX'$. Hence, $\angle X'XA = \Pi(XY) = \Pi(YA) = \angle MAX$. That is, XX' and AM are equally inclined toward AX, so $XX'|||\simeq AM$, as was to be shown. ∎

The converse proposition is essentially a repetition of the last paragraph of the preceding proof.

Claim 13. Lobachevski's horocycles are L-curves.

Proof. (See the preceding figure.) Let $\mathcal{H}$ be a horocycle, and let AM be one of its axes. We must show that if X is any point on $\mathcal{H}$, then $XN ||| \simeq AM$ for some ray XX'. Let XX' be the axis of $\mathcal{H}$ emanating from X. As axes of the horocycle, XX' and AM are parallel. To show that they are equally inclined toward AX, let YY' be the perpendicular bisector of AB. By definition of a horocycle, YY' is parallel to the axes XX' and AM. If we let $d = AC = AB$, then it is clear that $\angle X'XY = \Pi(d) = \angle MAY$.

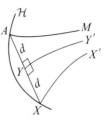

Thus, $XX' ||| \simeq AM$, as claimed. Hence, $\mathcal{H}$ lies within an L-curve, L. That is, $\mathcal{H}$ lies within a horocycle L (since we have just proved that L-curves are horocycles). Since one horocycle cannot contain another as a proper subset, we have $\mathcal{H} = L$. That is, $\mathcal{H}$ is an L-curve, as claimed. ∎

Gauss' Corresponding Points

In a brief unpublished note on parallels[2], Gauss nearly defined the horocycle. Given two parallel lines AA' and BB', Gauss defined A and B to be *corresponding points* when AB is equally inclined toward AA' and BB'. Since A and B "correspond" precisely when $AA' ||| \simeq BB'$, it is easy to reformulate Bolyai's definition of the horocycle using Gauss' terminology. Given a pencil of parallel lines (*i.e.* the family of all lines parallel to a given line), let us amplify Gauss' definition slightly and say that two points *correspond with respect to the pencil* if the line that joins them is equally inclined toward the lines of the pencil passing through them. Furthermore, let us adopt the convention that every point corresponds with itself. We may now define a horocycle as *the set of all points that correspond to some point A with respect to some pencil of parallels.*

This definition is more concise than either Lobachevski's or Bolyai's, but we should not be too hasty in praising Gauss for creating it. It is certainly implicit in his definition of corresponding points, but Gauss never actually wrote it down; he merely defined corresponding points and listed three facts about them without proofs.

In an oft-quoted letter of March 6, 1832 to Farkas Bolyai, Gauss wrote of János' *Appendix*, "to praise it would amount to praising myself: for the entire contents of the work, the path which your son has taken, the results to which he is led, coincide almost exactly with my own meditations which have occupied my mind for from thirty to thirty-five years. On this account I find myself surprised to the extreme."[3] After proceeding to offer what János would bitterly describe, twenty years later, as "pious wishes and complaints about the lack of adequate civilization,"[4] Gauss went on to suggest that Bolyai replace the "naked symbols" in his work, such as L and F, with descriptive names such as *paracycle* and *parasphere* respectively—names that he, Gauss, had thought of long ago.

---

[2] Gauss, p. 207.

[3] Gauss, pp. 220–221. Reb Hastrev has written a poem (unpublished) that includes the apposite verses, "And what of János Bolyai, who/With penetrating logic drew/Conclusions of profound degree/'Praise him?' Gauss cried, 'I'll first praise me!' "

[4] Greenberg, p. 142.

It would seem then that Gauss was fully aware of the horocycle's importance in the new geometry, but the credence that we should give Gauss' claims for priority has long been a matter of debate, exacerbated by the fragmentary nature of the evidence. A recent (2004) overview of this vexed question can be found in Jeremy Gray's appendix to G.W. Dunnington's biography of Gauss.[5] Had Gauss already followed the tortuous path from corresponding points to non-Euclidean trigonometry (via horocycle and horosphere) before reading Bolyai's work? Perhaps so. On the other hand, the path is neither easy to find nor easy to traverse, and even the great Gauss might have passed it by without exploring it thoroughly.[6]

Gauss' writings mention corresponding points in only one other place—near the end of a terse list of nine items[7] under the heading, "Parallelismus". The first five encompass the definition of, and basic statements about, parallels. The sixth reads: "What corresponding points on two parallel lines are." The seventh and eighth items assert properties of corresponding points, and the final cryptic entry reads: "*Trope ist die L*". Paul Stäckel, who compiled Gauss' unpublished notes, seems to have interpolated Gauss' probable meaning in brackets for their publication in Gauss' complete works, where the ninth item on the list appears as, "9. *Trope ist die L[inie, die von correspondirenden Punkten gebildet wird, wenn man alle Parallelen zu einer Geraden betrachtet.]*" ("*Trope* is the L[ine formed by corresponding points. . .]") Stäckel suggests that Gauss compiled the list in 1831, prompting Bonola to observe, "It is interesting to notice that Gauss, even at this date, seems to have anticipated the importance of the Horocycle. The definition of Corresponding Points and the statement of their properties is evidently meant to form an introduction to the discussion of the properties of this curve, to which he seems to have given the name *Trope*."[8] I propose a simpler explanation: Gauss drew up this undated list *after* reading Bolyai's work in 1832. In this case, *die L* would simply refer to Bolyai's *L*-curve, with no interpolation needed. Moreover, it seems more probable that Gauss would have switched his allegiance from *Paracycle* to *Trope* at some point *after* his letter to Bolyai rather than just before it, since that letter indicates that Bolyai's *L* had been known privately to Gauss as the *paracycle* for many years ("*vor langer Zeit*").

---

[5] Dunnington, pp. 461–7.

[6] The one note he left on non-Euclidean trigonometry was written after he had read both Bolyai and Lobachevski. Significantly, this note was discovered inside of his copy of Lobachevski's *Theory of Parallels*. (Dunnington, p. 186).

[7] Gauss, pp. 208–9.

[8] Bonola, p. 74, footnote.

Theory of Parallels 32

A circle of increasing radius merges into a horocycle.

When Lobachevski introduced the horocycle in TP 31 with the words, "*Grenzlinie (Oricycle) nennen wir ...*" (*We shall define a boundary-line (horocycle) ...*), he offered two names for it: the now familiar "horocycle", with its suggestions of circle-like properties, and *Grenzlinie*, meaning "boundary-line" or "limiting curve". In his final exposition of the subject, *Pangeometrie* (1855), he combined the circle and limit imagery into a single French name, *cercle limite*.[1]

"Horocycle" has become the standard term, and I have taken the translator's liberty of making exclusive use of it, despite the fact that Lobachevski favors *Grenzlinie* in the German original. Regardless of which name one ultimately settles on, it remains important to understand why the alternatives are reasonable. Explaining the sense in which the horocycle is a "limiting curve" or "limit circle" was, in fact, Lobachevski's sole purpose for including TP 32 in the present work; its inclusion was not strictly necessary from a logical standpoint, as TP 32 is never used in any subsequent proposition.

Since Lobachevski takes no pains to explain what he means by his phrase "*a circle of increasing radius merges into a horocycle*," I shall devote a few words to the intuitive meaning of this statement before examining its proof.

The Intuitive Picture

Let AC be a ray in *Euclidean* geometry, and l the perpendicular to it that passes through A. Let any point E on the ray determine a circle with center E and radius EA. As E slides down the ray, its corresponding circle grows. Imagine standing at a fixed spot on ray AC, perhaps a few feet away from A, and looking at line l. Let E be a point between yourself and A. This point slides toward you; its corresponding circle grows larger. At first, you are outside of the circle and can witness the whole shape growing in front of you, but

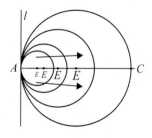

it soon overtakes you. You find yourself within its circumference. Point E passes under your feet and continues to recede down ray AC. Most of the circle is now behind your back, and hence out of your field of vision. Time passes, and the arc of the circle that remains in sight as you

---

[1] Actually, the two images are already combined in "horocycle", though few would recognize it. Most English words with the *hor-* prefix, such as "horoscope" or "horology", derive from the Latin *hora*, meaning "hour", "season", or "time". A few, however, such as *horizon*, come from the Greek *horos*, meaning "boundary." Lobachevski's "horocycle" belongs to this category.

look towards *l* becomes ever flatter. Eventually, you cannot distinguish it from *l*. Curious, you walk toward *l* to look closer. You find that at points near *A*, the circle is indistinguishable from *l*, as far as human eyesight can discern. Is this the case all over *l*, or just near *A*? You walk along *l*, as if walking up the beach, to investigate. Eventually, you reach a point at which you can discern a space between the still-growing circle and the line, but on closer examination, you find that the gap is shrinking; after a few minutes, it shrinks away to imperceptibility. You walk further along *l* to find a spot where the circle has yet to catch up with the line. You find one, but here too the gap vanishes as the circle's radius increases. No matter how far you walk up *l*, the circle eventually becomes indistinguishable from it. In this sense, the circle of increasing radius in Euclidean geometry "merges into" line *l*.

A bit more precisely, but still picturesquely, we can describe this merging situation as follows: if we walk along *l* for a while, stop at an arbitrary point *B*, and then turn 90° to look straight away from *l* toward the growing circle, then the point *F* at which the increasing circle intersects our line of sight will approach *B* as the radius of the circle increases.

Given the same scenario in imaginary geometry, the circle of increasing radius will *not* merge into *l*. Instead, TP 32 tells us that the circle will merge into the *horocycle* $\mathcal{H}$ whose axis is *AC*. Lobachevski's proof is essentially a verification of the property described in the preceding paragraph, substituting $\mathcal{H}$ for *l*. That is, he proves TP 32 by verifying the following: If *B* is an arbitrary point on $\mathcal{H}$, and *F* is the point at which the line erected perpendicular to $\mathcal{H}$ at *B* meets the growing circle, then $F \rightarrow B$ as the circle's radius increases. Of course, the line erected perpendicular to $\mathcal{H}$ at *B* is none other than the axis of $\mathcal{H}$ passing through *B* (see TP 31 Notes, Claim 10).

Before we examine Lobachevski's proof in detail, we make the following important observation: Regardless of how large the circle grows, it will never touch the horocycle at any point other than *A*. (Proof: If the two curves could share another point *X*, then they would have a common chord, the line segment *AX*. Consider the perpendicular bisector of such a chord. It would be parallel to axis *AC*, by definition of a horocycle. On the other hand, Euclid III.1 implies that it would cut *AC* at the center of the circle. Contradiction.)

We turn now to the details of Lobachevski's proof.

The Proof

Let *AB* be a chord of the horocycle. From its endpoints, *A* and *B*, draw the two axes *AC* and *BD*; these will necessarily make equal angles, $BAC = ABD = \alpha$, with the chord *AB* (TP 31). From either axis, say *AC*, select an arbitrary point *E* to be the center of a circle. Draw an arc of this circle extending from *A* to *F*, the point at which it intersects *BD*. The circle's radius *EF* will make angle $AFE = \beta$ on one side of the chord of the circle, *AF*; on the other side, it will make angle $EFD = \gamma$ with the axis *BD*.

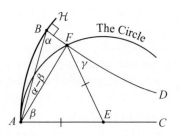

These opening sentences merely set the scene that we have already described. In particular, axis *BD* is the perpendicular to the horocycle erected at *B*. The equality $\angle BAC = \angle ABD$ follows from the fact that any two axes of a horocycle $\mathcal{H}$ are equally inclined toward the chord joining the points at which they intersect $\mathcal{H}$. (Recall that this property is part of Bolyai's very definition of the horocycle.)

It follows that the angle BAF between the horocycle's chord and the circle's chord is $BAF = \alpha - \beta < \beta + \gamma - \alpha$. From this, it follows that $\alpha - \beta < \frac{1}{2}\gamma$.

To establish TP 32, Lobachevski must show that $F \to B$ (where F is defined as the intersection of the growing circle and the axis BD) as the circle's radius increases. Showing that $F \to B$ is clearly equivalent to showing that $\angle BAF \to 0$. Lobachevski establishes this latter limit. His first step in this direction is to show that $\angle BAF < \frac{1}{2}\gamma$. He will then finish the proof by demonstrating that $\gamma \to 0$ as the circle grows. We may establish his preliminary inequality as follows:

$$\angle BAF = \angle BAE - \angle FAE = \alpha - \angle FAE$$
$$= \alpha - \angle AFE \quad \text{(The base angles of isosceles triangle } \triangle AFE \text{ are equal.)}$$
$$= \alpha - \beta.$$

Then, since the sum of the angles in $\triangle ABF$ is less than π, we have

$$\angle BAF + \angle AFB + \angle ABF < \pi.$$

That is,

$$(\alpha - \beta) + (\pi - \beta - \gamma) + \alpha < \pi.$$

Equivalently,

$$(\alpha - \beta) < \pi - (\pi - \beta - \gamma) - \alpha.$$

That is,

$$(\alpha - \beta) < \gamma - (\alpha - \beta),$$

from which it follows that

$$(\alpha - \beta) < \frac{1}{2}\gamma, \text{ as claimed.}$$

Having secured the inequality $\angle BAF < \frac{1}{2}\gamma$, it remains for Lobachevski to show that $\gamma \to 0$ as the circle grows. He addresses this point next.

Now, angle γ will decrease if we move F toward B along axis BF while holding the center E fixed (TP 21). Moreover, γ will decrease to zero if we move the center E down axis AC while holding F fixed (TP 21, 22).

As the circle grows, point E and point F both move. In turn, their motions alter γ, the measure of $\angle EFD$. Lobachevski's argument that γ vanishes as the circle becomes infinitely large is somewhat objectionable: although the two variables (the locations of E and F) are *not* independent of one another, he analyzes them as though they were. We can fix this as follows.

Suppose that, before it begins to grow, the circle is *initially* centered at E' and *initially* intersects BD at F', as shown in the figure. Then, as the circle grows, its moving center E slides down ray $E'C$ while F slides up ray $F'B$. The motion of F makes it clear that $\angle AEF$ will always be less than $\angle AEF'$. Since the latter angle vanishes as the circle becomes infinitely large (by TP 21), it follows that the former

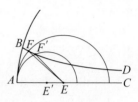

angle must do the same. Finally, to conclude this part of the argument, we shall show that γ, the measure of $\angle EFD$, is always less than the vanishing quantity $\angle AEF$, and hence must itself vanish.

Claim. $\gamma < \angle AEF$.

Proof. We shall show that the alternatives lead to contradictions.

First, suppose $\gamma = \angle AEF$. Bisect EF at M. By dropping perpendiculars MP and MQ to BD and AC respectively, we produce congruent triangles $\triangle FMP \cong \triangle EMQ$ (by AAS). Hence, $\angle FMP = \angle EMQ$. Let θ be the common measure of these angles. Then $\angle PME$, the supplement of $\angle FMP$, is $\pi - \theta$, so that $\angle PMQ = \angle PME + \angle EMQ = \pi - \theta) + \theta = \pi$. That is, P, M, and Q are collinear. Consequently, line PQ is a common perpendicular to the parallels BD and AC, which is impossible by TP 22. Hence, $\angle AEF \neq \gamma$.

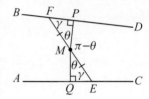

Next, suppose $\gamma > \angle AEF$. Draw FG such that $\angle EFG = \angle AEF$. Because $FD\|AC$, line FG must cut AC (TP 16) at some point H. Then the exterior angle $\angle AEF$ of $\triangle EFH$ is equal to $\angle EFH$, one of its remote interior angles, contradicting Euclid I.16. Hence, $\angle AEF$ is not less than γ.

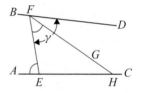

Having exhausted the alternatives, we conclude that $\gamma < \angle AEF$, as claimed. ∎

Thus, we have demonstrated that γ vanishes as the growing circle becomes infinitely large.

As γ vanishes, so does $\alpha - \beta$, the angle between AB and AF. Consequently, the distance from point B of the horocycle to point F of the circle vanishes as well.

We have already seen that $(\alpha - \beta) < \frac{1}{2}\gamma$. Hence, the growth of the circle causes $(\alpha - \beta)$ to vanish along with γ. That is, $\angle BAF$ vanishes as the circle grows. Since only one arm of this angle, AF, moves in response to the circle's growth, it must approach the other arm, AB, if the angle is to vanish. Thus, $F \to B$, as claimed, and the circle of increasing radius merges into the horocycle in the sense described above.

For this reason, one may also call the horocycle a circle of infinite radius.

The circle approaches the horocycle (its "limiting curve") as its radius increases toward infinity. Thus, by indulging in the traditional "abuse of language" associated with limiting behavior, we may say that when the circle's radius *actually is* infinite, the circle *actually is* the horocycle. That is, we may view the horocycle as a "circle of infinite radius".

Circle-like Properties of the Horocycle: A Summary

Here are five of the most vital facts supporting the interpretation of a horocycle as a circle whose center is a point at infinity, and whose "diameters" are its axes.

1. "A circle of increasing radius merges into a horocycle." (TP 32)
2. The perpendicular bisectors of a horocycle's chords all "meet at a point at infinity". That is, they are parallel to one another. Circles share this property, except that the bisectors meet at an ordinary point, not at infinity. (TP 31, Lobachevski's definition of a horocycle)
3. A horocycle is symmetric about its axes, just as a circle is symmetric about its diameters. (TP 31 notes, Claim 4)
4. A horocycle is orthogonal to its axes, just as a circle is orthogonal to its diameters. (TP 31 notes, Claim 10, Corollary)
5. A horocycle is determined by a point at infinity (its center) and one ordinary point (a point on its "circumference"), just as a circle is determined by its center and one point on its circumference. (TP 31 notes, Claim 9)

Two Line-like Properties

Two attributes of horocycles, however, mark them as peculiarly line-like.

1. All horocycles are congruent to one another.
2. In the presence of the parallel postulate, a horocycle *is* a straight line.

Because Euclid constructs all of the geometric figures in *The Elements* with straightedge and compass, the tension in his work derives, in one sense, from the seemingly contrary natures of lines and circles—the very exemplars of perfect straightness and uniform curvature. One wonders what he would have thought of a curve in which those natures intertwine.

Theory of Parallels 33

Let $AA' = BB' = x$ be segments of two lines that are parallel in the direction from A to A'. If these parallels are axes of two horocycles, whose arcs $AB = s$ and $A'B' = s'$ they delimit, then the equation $s' = se^{-x}$ holds, where e is some number independent of the arcs s, s', and the line segment x, the distance between the arcs s' and s.

To simplify the somewhat confusing statement of TP 33, we shall introduce some new terminology.

Preliminaries: Concentric Horocycles

'It's time for a definition,' he said.
'Then follow my lead,' I replied, 'and we'll see if we can reach a satisfactory explanation somehow or other.'
—Plato, *Republic*, 474c

Recall that a horocycle is determined by two data: a point upon it, and its center (intuitively, the center is a point at infinity; formally, it is a pencil of parallels—the set of the horocycle's axes). Naturally, *concentric horocycles* are defined to be horocycles sharing the same center. In other words, two horocycles are concentric if and only if their sets of axes are identical. We define the *distance between two concentric horocycles* to be the length of any axis cut off between them; this length does not depend upon the particular axis that we choose to measure, as we now demonstrate.

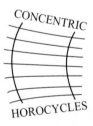

Claim 1. The distance between two concentric horocycles is a well-defined concept, inasmuch as it does not depend upon the axis we choose to measure.

Proof. Let AA' and BB' be segments of axes cut off by the same pair of concentric horocycles, as in the figure.

We must prove that they have the same length.

Every horocycle satisfies the definition of Bolyai's L-curve (TP 31 notes, Claim 13); that is, every horocycle has the property that any two of its axes are equally inclined toward the chord joining their endpoints.

Consequently, $\angle B'BA = \angle A'AB$ and $\angle AA'B' = \angle BB'A'$.

Erect a perpendicular to AB at its midpoint M.

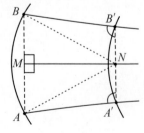

Let N be the perpendicular's intersection with $A'B'$.

Draw AN and BN.

We immediately have $\triangle NAM \cong \triangle NBM$ (by SAS). From this, it follows that $NA = NB$ and $\angle NAM = \angle NBM$. Subtracting equals from equals, we have $(\angle B'BA - \angle NBM) = (\angle A'AB - \angle NAM)$. That is, $\angle A'AN = \angle B'BN$.

Consequently, $\triangle A'AN \cong \triangle B'BN$ (by AAS).

Hence, $AA' = BB'$, as claimed. ∎

TP 33 is a theorem about concentric horocycles. Once we recognize this, we can reformulate the statement of the theorem as follows.

TP 33 (Rephrased). *Consider two concentric horocycles and two of their common axes, AA' and BB'. If x is the distance between the horocycles, while s and s' are the lengths of their arcs which lie between the axes, as shown in the figure, then the three lengths s, s', and x will be related by the formula*

$$s' = se^{-x},$$

where e is a constant whose numerical value is determined by the unit with which we measure length: by a suitable choice of unit, we can endow e with any value (greater than 1) that we please. (Lobachevski eventually chooses the unit of length so that the value of e is the base of the natural logarithm.)

The most curious feature of this formula is the constant e, whose numerical value depends upon the size of our measuring stick. Such constants are unknown in Euclidean geometry, but they are actually quite common in *spherical* geometry. Consider, for example, how a spherical triangle's area and angular excess are related. If our unit is one-fifth of the sphere's radius, then the radius is 5 units long, and the area of a spherical triangle is given by $A = 25 \times excess$. In contrast, if we take the sphere's diameter as our unit of length, then the radius is $\frac{1}{2}$ a unit long, and the area of a spherical triangle is given by $A = \frac{1}{4} \times excess$. In general, we have $A = r^2 \times excess$, where r is a constant whose numerical value is determined by the unit with which we measure length. Similarly, the spherical Pythagorean theorem is $\cos(c/r) = \cos(a/r)\cos(b/r)$, where r's numerical value (the length of the sphere's radius) depends upon our unit of measurement.

The existence of a constant whose numerical value is determined by the unit of length is therefore not without precedent. However, the comfort afforded by this apparent similarity to the tangible world of spherical geometry begins to wear thin when we reflect upon a crucial difference: spherical geometry's ubiquitous r has a natural interpretation as the sphere's radius, but the indeterminate constant of imaginary geometry has no clear geometric significance.

This unsettling aspect of imaginary geometry may have actually been responsible for the lengthy delay in the publication of János Bolyai's work. On November 3, 1823, he wrote to his father, Farkas Bolyai, that he had "created a new and different world out of nothing". In his response, Farkas urged his son to publish his results quickly, arguing that "since all scientific striving is only a great war and one does not know when it will be replaced by peace, one must win, if possible; for here, preeminence comes to him who is first." His words proved prophetic. When János visited his father in February 1825 to discuss his "new world", hoping that his father would help him get it into print, he was disappointed to find that, in the words of Jeremy Gray, "he was unable to convince him, worried as he was about an arbitrary constant that entered the formulae his son had found."[1] Consequently, Bolyai's work was not published until 1832. During the long delay, Lobachevski became (in 1829) the first man to publish an account of non-Euclidean geometry.

A Missing Lemma

We shall need the following plausible result, which Lobachevski assumes without proof.

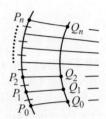

Claim 2. ("Equal Division Lemma") Let $\mathcal{H}$ and $\mathcal{K}$ be concentric horocycles. If $P_0, P_1, \ldots, P_n$ are equally spaced points on $\mathcal{H}$ (so that the horocyclic arcs $P_0P_1, P_1P_2, \ldots, P_{n-1}P_n$ all have the same length), then the axes passing through them meet $\mathcal{K}$ in points $Q_0, Q_1, \ldots, Q_n$, which are equally spaced as well.

Proof. The symmetry of horocycles about their axes has the following consequence:

For any distinct points A, B, and C of a horocycle, the arcs AC and CB have the same length if and only if A and B are swapped by reflection in the axis passing through C.

We will use this fact to prove our lemma. To begin, we shall show that $arc Q_0Q_1 = arc Q_1Q_2$.

By hypothesis, $arc P_0P_1 = arc P_1P_2$. Thus, P_0 and P_2 are swapped by a reflection in axis P_1Q_1.

Hence, the image of line P_0Q_0 under this reflection is a line passing through P_2.

Moreover, since reflection preserves parallelism, the image of P_0Q_0 must be parallel to the image of P_1Q_1, which is P_1Q_1 itself. Thus, the reflected image of P_0Q_0 is a line passing through P_2 which is parallel to P_1Q_1. The only such line is P_2Q_2. Hence, the axes P_0Q_0 and P_2Q_2 are swapped by the reflection.

Consequently, points Q_0 and Q_2 are swapped by the reflection in axis P_1Q_1. Hence, $arc Q_0Q_1 = arc Q_1Q_2$.

Repeating the argument, but using reflection in axis P_2Q_2, we can show that $arc Q_1Q_2 = arc Q_2Q_3$.

Thus, $arc Q_0Q_1 = arc Q_1Q_2 = arc Q_2Q_3$.

Continuing in this fashion, we can show that $arc Q_0Q_1 = arc Q_1Q_2 = arc Q_2Q_3 = \cdots = arc Q_{n-1}Q_n$. That is, the points $Q_0, Q_1, \ldots, Q_n$ are equally spaced along horocycle $\mathcal{K}$, as claimed. ∎

Overview of the Proof of TP 33

Before diving into the details of the proof, we shall examine the broad outline of the argument, which essentially falls into two steps.

---

[1] Gray, *János Bolyai.* pp. 52–3.

Step 1 (Definition and arc-invariance of *the shrinking factor*). To *project* an arc of a horocycle onto an interior concentric horocycle, we simply "slide it" down the axes common to the two curves. (For example, in the figure at right, arc AB of $\mathcal{H}$ projects onto arc $A'B'$ of $\mathcal{K}$.) Accordingly, given a pair of concentric horocycles, any arc of the exterior one (such as AB in the figure) determines two arclengths: its own (which we shall call s), and the arclength of its projection onto the interior horocycle (which we shall call s').

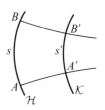

Individually, the numerical values of s and s' depend upon the unit of measurement, but their *ratio* (say, s/s', the bigger to the smaller) does not; it is a pure, dimensionless number[2]. In fact, this ratio is unaffected even by our choice of the arc (AB) with which we determine s and s'. In other words, if I select a tiny arc on the outer horocycle and divide its length by the length of its projection, I will obtain the same value that you will, even if you select an enormous arc for the same procedure. Demonstrating this arc-invariance will be the first step of TP 33's proof.[3]

I shall call s/s', the invariant ratio of arclengths, *the expansion factor associated with the two concentric horocycles*. For, given an arc of the inner horocycle, when we multiply its length by the expansion factor, we will obtain the length of its projection onto the outer horocycle. Similarly, I shall call s'/s, the reciprocal ratio, *the shrinking factor associated with the two concentric horocycles*.

Step 2 (The shrinking factor varies exponentially with the distance between the horocycles). The expansion/shrinking factor clearly *does* depend upon the distance between the horocycles. If they are barely separated from one another, then projections will cause minimal shrinking or expanding. Because the axes of concentric horocycles draw ever closer to one another in the direction of their parallelism (TP 24), distantly separated horocycles will have more pronounced expansion/shrinking factors.

Since the shrinking factor is a function of the distance between the concentric horocycles with which it is associated, we may write $s'/s = f(x)$, for some function f, where x represents the distance between the two concentric horocycles (as measured by some fixed unit of length). Equivalently, $s' = s \cdot f(x)$, where $f(x)$ is the shrinking factor associated with concentric horocycles separated by distance x. Note that this is very close to Lobachevski's formula, $s' = se^{-x}$. Thus, the second step in the proof will be to show that $f(x)$ is an exponential function. That is, the shrinking factor varies exponentially with the distance between concentric horocycles. We shall do this by demonstrating that $f(x)$ satisfies a *functional equation* whose only solutions are exponential functions. This functional equation is well known, but I will record it here as a lemma before examining Lobachevski's proof.

A Functional Equation

Claim 3. If f is a continuous function such that $f(x + y) = f(x)f(y)$ for all positive reals x and y, and $f(0) = 1$,[4] then

---

[2] For example, if I use inches, I might find $s = 18$ and $s' = 9$, while you, using feet, would find $s = 1.5$ and $s' = 0.75$. Although we would disagree on the numerical values of s and s', we would agree that $s/s' = 2$.

[3] Since this type of arc-invariance obviously holds when we carry out the same procedure on concentric *circles*, we should not be too surprised to meet it in the context of horocycles, which we think of, after all, as circles of infinite radius.

[4] In fact, $f(0) \neq 0$ will suffice for this proof, but we will use this lemma only in cases where $f(0) = 1$.

$$f(x) = a^x, \text{ where } a = f(1).$$

Proof. Let $a = f(1)$. For any natural number n, we have $f(n) = f(1 + 1 + \cdots + 1) = f(1)f(1) \cdots f(1) = [f(1)]^n = a^n$.

Thus, for any natural numbers n and m, we have $a^n = f(n) = f(n/m + \cdots + n/m) = f(n/m)^m$.

Hence, $f(n/m) = a^{n/m}$. That is, $f(x) = a^x$ for all positive rational values of x.

Since the rationals are dense in the positive reals, f's continuity guarantees that $f(x) = a^x$ for all positive real values of x. ∎

The Proof: Step 1

Suppose that n and m are whole numbers such that $s : s' = n : m$. Draw a third axis CC' between AA' and BB'. Let $t = AC$ and $t' = A'C'$ be the lengths of the arcs that it cuts from AB and $A'B'$ respectively. Assuming that $t : s = p : q$ for some whole numbers p and q, we have

$$s = (n/m)s' \text{ and } t = (p/q)s.$$

If we divide s into nq equal parts by axes, any one such part will fit exactly mq times into s' and exactly np times into t. At the same time, the axes dividing s into nq equal parts divide s' into nq equal parts as well. From this it follows that

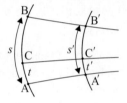

$$t'/t = s'/s.$$

Consequently, as soon as the distance x between the horocycles is given, the ratio of t to t' is determined; this ratio remains the same, no matter where we draw CC' between AA' and BB'.

As described in the "overview", the arc AB determines the arclengths s and s'. (See the figure above.) We must show that the ratio $s' : s$, the shrinking factor, is independent of the arc AB. To this end, we take a second arc at random (Lobachevski uses AC),[5] denote its arclength by t, and denote the length of the arc it determines on the second horocycle by t'. We must prove that $t' : t = s' : s$.

Lobachevski's argument implicitly involves four cases. He explicitly proves the first case, in which s is commensurable with both s' and t, but trusts his readers to supply the details of the remaining three cases (in which s is incommensurable with s', incommensurable with t, or incommensurable with both s' and t.) In fact, Lobachevski's argument is more involved than necessary. Whether s is commensurable with t is an important distinction; whether s is commensurable with s' is irrelevant, as I shall now demonstrate, by proving that $t' : t = s' : s$ in only two cases.

---

[5] Admittedly, only *one* endpoint of AC is random, but once we prove that the ratio is the same for arcs with one endpoint in common, we can easily extend this to a completely random arc as follows: let GH be any arc whatsoever. The ratio is the same for AB and AH, since they have A in common; similarly, the ratio is equal for AH and GH, since they have H in common. Thus, the ratio is identical for AB and the random arc GH.

Case 1 (*s* and *t* are commensurable). In this case, there are whole numbers *p* and *q* such that $s = qu$ and $t = pu$.

Divide AB into q equal arcs of length *u*, the first *p* of which divide AC evenly. By the equal division lemma (Claim 2), the axes passing through the points of division cut $A'B'$ into q equal arcs, the first *p* of which divide $A'C'$ evenly. Denoting the common length of these arcs by *u'*, we have that $s' = A'B' = qu'$, and $t' = A'C' = pu'$.

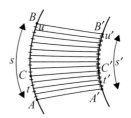

Therefore, $t' : t = pu' : pu = u' : u = qu' : qu = s' : s$.

That is, $t' : t = s' : s$, as claimed.

Case 2 (*s* and *t* are incommensurable). Divide AC into *n* equal arcs. Let $u = t/n$ denote their length. Begin at *A* and measure off arcs of length *u* along AB until we reach a point *X* (before *B*) such that the arc XB has length less than $u = t/n$. Since *u* measures both AC and AX, these arcs are commensurable, and thus, by Case 1, we have $A'C' : AC = A'X' : AX$.

Now, when $n \to \infty$, we have that $(u = t/n) \to 0$. Thus, $X \to B$, so that $A'X' : AX \to A'B' : AB$. On the other hand, $A'X' : AX$ is always equal to $A'C' : AC$, which is a constant, so we must have $A'X' : AX \to A'C' : AC$.

These two expressions for the same limit must be equal, so we have

$$A'B' : AB = A'C' : AC.$$

That is, $t' : t = s' : s$, as claimed.

Thus, we have shown that $t' : t = s' : s$ in any case. That is, the shrinking factor is arc-independent, as claimed. This completes this first step of the proof. As discussed in the overview, this implies that $s' = s \cdot f(x)$ for some function $f(x)$, which represents the shrinking factor as a function of the distance between two concentric horocycles.

The Proof: Step 2 (A se^{-x} y Formula)

From this, it follows that if we write $s = es'$ when $x = 1$, then $s' = se^{-x}$ for every value of *x*.

When Lobachevski writes $s = es'$ when $x = 1$, he is effectively *defining* the symbol *e* as the expansion factor for concentric horocycles separated by one unit of distance. (It may be helpful to think of *e* as shorthand for "*e*xpansion" in this context.) It is crucial to understand that *e*, at this point in the argument, has nothing to do with the base of the natural logarithm.

Thinking of *e* as an expansion factor reveals why its numerical value depends upon our unit of length. If our unit of length is the millimeter, then *e* is the expansion factor between nearly coincident horocycles. Clearly, it will be very close to 1 in this case. On the other hand, if we use the light-year as our unit of length, then *e* may be considerably larger than 1. We shall return to the numerical value of *e* shortly. First, let us prove that $f(x)$, the shrinking factor for concentric horocycles separated by distance *x*, is an exponential function.

Claim 4. Consider two concentric horocycles separated by a distance of *x*. If *s* and *s'* are corresponding arcs on the horocycles (*s* on the outer, *s'* on the inner), then $s' = se^{-x}$, where *e* is the expansion factor for concentric horocycles separated by a unit distance. (The numerical value of e depends on the unit of length.)

Proof. The figure at right shows at a glance that $f(x + y) =$ $f(x)f(y)$. The outer and inner horocycles are separated by $x + y$, so an arc of length 1 on the outer projects to an arc of length $f(x + y)$ on the inner. Alternatively, the outer arc projects to an arc of length $f(x)$ on the middle horocycle, which in turn projects to an arc on the inner horocycle of length $f(x)f(y)$. If we equate the two expressions for the innermost horocyclic arc, we obtain $f(x + y) = f(x)f(y)$, so by Claim 3, $f(x) = f(1)^x$.

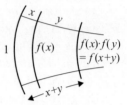

Since $f(1)$ is the shrinking factor between concentric horocycles separated by one unit, it is the reciprocal of e, the expansion factor between such horocycles. That is $f(1) = e^{-1}$. Hence, $f(x) = (e^{-1})^x = e^{-x}$. Since $s' = sf(x)$, as we saw in Step 1 above, it follows that $s' = se^{-x}$, as claimed. ∎

Corollary. *Parallel lines are asymptotic (i.e. the distance between them not only decreases, but decreases to zero in their direction of parallelism.)*

Proof. Regardless of the unit of length, we know that $e > 1$ since e is an expansion factor. Hence, the formula $s' = se^{-x}$ implies that $s' \to 0$ as $x \to \infty$. ∎

It remains only to clarify the meaning of Lobachevski's e and relate it to the base of the natural logarithm.

A Unit of Length

We may choose the unit of length with which we measure x as we see fit. In fact, because e is an undetermined number subject only to the condition $e > 1$, we may, for the sake of computational ease, choose the unit of length so that the number e will be the base of the natural logarithm.

In addition, since $s' = 0$ when $x = \infty$, we observe that, in the direction of parallelism, the distance between two parallels not only decreases (TP 24), but ultimately vanishes. Thus, parallel lines have the character of asymptotes.

We have tacitly assumed that we have a unit of length with which to measure x, s, and s', but we have never bothered to describe it in any detail. Above, we saw that the choice of unit determines the numerical value of the expansion factor, e. Lobachevski proposes that we reverse this process and let a number determine our unit of length.

Specifically, we may define our unit of length as follows. Let $\mathcal{H}$ be a fixed horocycle, and let $\mathcal{K}$ be a second horocycle superimposed on top of it. Move $\mathcal{K}$ away from $\mathcal{H}$ so that the horocycles remain concentric as the distance between them increases. The expansion factor associated with the horocycles is initially 1 (when they are coincident), but it increases as they separate. In fact, because the axes are asymptotic in their direction of parallelism, the expansion factor increases without bound as the distance between the horocycles increases. Hence, at some point, the expansion factor will be precisely the base of the natural logarithm ($2.71828\ldots$). At this moment, we stop moving $\mathcal{K}$ and take the distance separating the two horocycles to be our unit of length.

With this carefully constructed unit of length, the expansion factor e in the formula $s' = se^{-x}$ is in fact the base of the natural logarithm.

Notation Variation

Lobachevski's use of e to represent an indeterminate constant is regrettable. Once we decide to arrange matters so that the e in $s' = se^{-x}$ is the familiar logarithmic base, we are apt to forget that we could have endowed it with any other value (greater than 1, of course) by choosing a different unit of length.

Lobachevski could have avoided this potential confusion by calling his expansion factor $e^{1/k}$, letting e retain its usual meaning and allowing the parameter k to take any positive value. This works because as k varies over the positive reals, $e^{1/k}$ assumes all values greater than 1, as the expansion factor should. This convention is neater, since it relates the unit length to the value of the parameter k, rather than to the numerical value of e. Viewed from this perspective, the formula in this proposition becomes, in its most general form,

$$s' = se^{-x/k}.$$

Adopting Lobachevski's choice of unit length is equivalent to setting $k = 1$. If we retain the parameter k instead of fixing its value at 1, we may express subsequent equations in more general terms. In future propositions, I shall include footnotes that indicate the form that Lobachevski's equations would take if he had retained the parameter. The extra generality manifests itself only in trivial changes in formulae; no new ideas are involved. On the other hand, seeing the more general forms reinforces the analogy with spherical geometry, whose formulae involve a parameter, r, whose value depends upon the unit of length.

Asymptotic Triangles

> "One of the most elegant passages in the literature on hyperbolic geometry since the time of Lobachevsky is the proof by Liebmann that the area of a triangle remains finite when all its sides are infinite."
>
> — H.S.M. Coxeter[6]

The convention that parallel lines meet one another at an ideal point at infinity seems particularly apt in light of the asymptotic nature of parallels. It also justifies the notion of an "asymptotic triangle": a triangle with at least one vertex at infinity, where two of its sides (two *infinitely long*, parallel sides) meet one another. Such triangles are called singly, doubly, or triply asymptotic, according to the number of its vertices they have at infinity.

We name ordinary triangles by listing their vertices (such as $\triangle ABC$); by denoting ideal vertices with capital omegas (Ω, Ω', Ω'', etc.), we can name asymptotic triangles in the same way. (See the figures.)

---

[6] Coxeter, p. 295

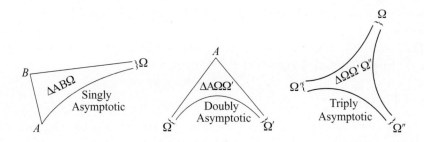

I shall indicate a few properties of asymptotic triangles, which we will need in order to understand a remarkable proof by Gauss that the area of an ordinary triangle in imaginary geometry is directly proportional to its angle defect. None of this material is needed to understand the remainder of *The Theory of Parallels*; I have included it simply because it is too elegant to ignore. Accordingly, some of the arguments that I present below are left deliberately sketchy, lest the details obscure the large ideas in this digression.

Claim 5. Given two rays AB and AC emanating from A, there exists a unique line which is parallel to AB in one direction and to AC in the other.

Proof. Let d be the unique length such that $\Pi(d) = \frac{1}{2}\angle BAC$ (such a d exists, by TP 23). Let D be the point on the angle bisector of $\angle BAC$ such that $AD = d$. Draw the line l through D which is perpendicular to AD. By construction, l will be parallel to AB on one side, and to AC on the other. Thus, a line of the sort that we desire *exists*. If m is another such line, then the transitivity of parallelism implies that l and m are parallel to one another *in both directions*, which is impossible (TP 24). Hence, l is the *unique* line parallel to AB in one direction, and AC in the other. ∎

If we think of the rays AB and AC in the preceding lemma as each "pointing to" a distinct point at infinity, we may interpret the lemma as an extension of Euclid's first postulate: *there is a unique line joining any two points at infinity*. Note that this line is perfectly ordinary; there is no need to postulate a "line at infinity", as we do in projective geometry.

Claim 6. The area of any singly asymptotic triangle is finite.

Proof. (sketch). Let $\triangle AB\Omega$ be a singly asymptotic triangle. Embed it in a doubly asymptotic triangle by extending its finite side AB to the *ray* AB, and drawing the unique line that is parallel to AB in one direction and parallel to $A\Omega$ in the other (claim 5). Letting Ω' denote the ideal vertex where this line "meets" AB, we have constructed a doubly asymptotic triangle $\triangle A\Omega\Omega'$. Let B' be the point on ray $A\Omega$ such that $AB' = AB$. From B and B', drop perpendiculars BC and $B'C'$ to $\Omega\Omega'$. Like any bounded figure, the pentagon $ABCC'B'$ has a finite area. By a dissection argument, one can show that the singly asymptotic triangle $\triangle AB\Omega$ has the same area as the pentagon. I shall omit the details of this argument[7], and merely present the "picture" of the proof (figure on the right) discovered by Heinrich Liebmann. ∎

---

[7] They may be found in Coxeter, pp. 295–6.

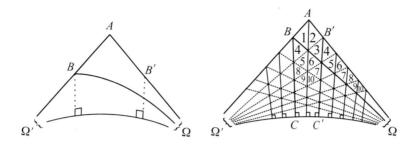

Corollary 1. *The area of any doubly asymptotic triangle is finite.*

Proof. By dropping a perpendicular from the ordinary (non-ideal) vertex of a doubly asymptotic triangle to its opposite side, we decompose it into two singly asymptotic triangles, each of which has finite area. The sum of their areas is the area of the doubly asymptotic triangle. ■

Corollary 2. *The area of any triply asymptotic triangle is finite.*

Proof. Choose a point on a side of a triply asymptotic triangle. From the point, draw a line parallel to the remaining sides. This decomposes the triply asymptotic triangle into two doubly asymptotic triangles. Its area is the sum of their areas, and so, by Corollary 1, is finite. ■

We typically describe an ordinary triangle by six data: the measures of its three angles, and the lengths of its three sides. However, each triangle congruence criterion requires the agreement of only three of these data (SAS, SSS, ASA, AAS, or AAA in imaginary geometry).

Singly asymptotic triangles have two infinite sides and an ideal vertex, and thus can vary in only three of the six data: the two ordinary angles, and their included side. Any two of these three will suffice to determine a singly asymptotic triangle up to congruence. Intuitively, this is because any two asymptotic triangles have equal 0° angles at their ideal vertices. If, in addition, they agree on two other data (AA or AS), we then have AAA or AAS, which implies that the triangles are congruent.

Similarly, a doubly asymptotic triangle can vary in only one datum: the angle at its sole non-ideal vertex. Any pair of doubly asymptotic triangles with equal angles will be congruent: they automatically agree on *two* 0° angles, so the equality of their remaining angles yields congruence by AAA.

Finally, *all* triply asymptotic triangles are congruent by AAA (where all angles are 0°).

This last fact is quite remarkable: like points, lines, and horocycles, any two triply asymptotic triangles are identical save for their location in the plane. Combined with Corollary 2, this tells us that *all triply asymptotic triangles have the same finite area.*

Gauss' Proof: Area and Defect are Proportional

> "... instead of expressing his great joy and interest, and trying to prepare an appropriate reception for the good cause, avoiding all these, he rested content with pious wishes and complaints about the lack of adequate civilization."
>
> — János Bolyai (1851), writing about Gauss' response to the *Appendix* in 1832.[8]

Gauss' 1832 letter to Farkas Bolyai concerning János' *Appendix* is perhaps best known for its infamous line, "to praise it would be to praise myself."[9] Considerably less well known is the beautiful proof Gauss sketched in this letter of the fact that the area of a triangle in imaginary geometry is directly proportional to its angle defect. The proof sketched in his letter is essentially a list of seven steps, culminating in the desired result.[10] I have amplified Gauss' argument in the proposition below, following a simple lemma.

Claim 7. If a continuous function f is *additive* (i.e. if $f(x + y) = f(x) + f(y)$), then $f(x) = kx$ for some real number k.

Proof. Let $k = f(1)$.
Case 1 (x is a natural number, n):

$$f(n) = f(1 + 1 + \cdots + 1) = f(1) + f(1) + \cdots + f(1) = k + k + \cdots + k = kn.$$

Case 2 (x is a positive unit fraction, $1/n$):

$$k = f(1) = f(1/n + 1/n \cdots + 1/n) = f(1/n) + f(1/n) + \cdots + f(1/n) = nf(1/n).$$

Hence, $f(1/n) = k/n$.
Case 3 (x is a positive rational number, m/n):

$$f(m/n) = f(1/n + \cdots + 1/n) = f(1/n) + \cdots + f(1/n) = m \cdot f(1/n) = m(k/n) = k(m/n).$$

Case 4 ($x = 0$):

$$f(0) = f(0 + 0) = f(0) + f(0).$$

Subtracting $f(0)$ from each side yields $f(0) = 0 = k0$.
Case 5 (x is a negative rational number, $-m/n$):

$$0 = f(0) = f(-m/n + m/n) = f(-m/n) + f(m/n).$$

Hence, $f(-m/n) = -f(m/n) = -k(m/n) = k(-m/n)$, as claimed.

---

[8] Greenberg, p. 142.

[9] Gauss does offer some relatively selfless praise to Bolyai in his letter ("I...am overjoyed that it happens to be the son of my old friend who outstrips me in such a remarkable way."), but one cannot help wishing that he had told him instead what he confided to his friend Gerling, "I consider this young geometer Bolyai to be a genius of the highest order." (Gauss, p. 220).

[10] Gauss, pp. 220–223.

We have now shown that $f(x) = kx$ holds for all rational values of x in the domain of f. As a result, it must also hold for irrational values of x as well, by the continuity of f. ∎

Claim 8. In imaginary geometry, the area of the triangle is directly proportional to its angle defect.

Proof. Let t be the finite area common to all triply asymptotic triangles.

The area of a *doubly* asymptotic triangle is a function of its sole non-zero angle, since this angle determines the triangle up to congruence. In this proof, it will be more convenient to express this area as a function of the *supplement* of this angle. Thus, if ϕ is the *external* angle of a doubly asymptotic triangle, we shall denote the triangle's area by $f(\phi)$. Note the extreme case $f(\pi) = t$. (If the external angle is π, then the internal angle is 0, in which case the triangle is triply asymptotic, and consequently has area t.)

We shall now show that f is an additive function. First, we establish additivity in two special cases.

Case 1. The figure below (left) shows that

$$f(\phi) + f(\pi - \phi) = t, \quad \text{for any} \quad \phi \le \pi. \tag{1}$$

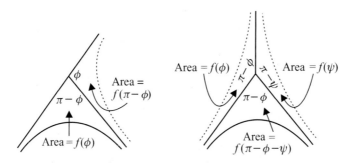

Case 2. The figure above (right) shows that

$$f(\phi) + f(\psi) + f(\pi - \phi - \psi) = t, \tag{2}$$

provided $\phi + \psi \le \pi$.

(To produce the right figure, we begin with a doubly asymptotic triangle $\triangle A\Omega\Omega'$ with angle $\phi + \psi$ at A. Draw ray $A\Omega''$ from A, forming angle $\pi - \phi$ with $A\Omega$. Then $\angle\Omega''A\Omega' = 2\pi - [(\pi - \phi) + (\phi + \psi)] = \pi - \psi$. Finally, Claim 5 allows us to draw the sides $\Omega\Omega''$ and $\Omega'\Omega''$ of the triply asymptotic triangle $\triangle\Omega\Omega'\Omega''$.)

By combining these two special cases, we can now establish additivity in general[11]:

---

[11] That is, we can establish that $f(\varphi) + f(\psi) = f(\varphi + \psi)$, for all values of φ and ψ, whose sum does not exceed π. This last condition reflects the geometric definition of the function f: its argument is a triangle's external angle, and hence must lie between 0 and π.

If $\phi + \psi \leq \pi$, then (2) gives $\qquad\qquad t = f(\phi) + f(\psi) + f(\pi - \phi - \psi)$.
Substituting $(\phi + \psi)$ for ϕ in (1) gives $\qquad t = f(\phi + \psi) + f(\pi - \phi - \psi)$.
Equating these two expressions for t yields $\qquad f(\phi) + f(\psi) = f(\phi + \psi)$.

Since f is additive, we know that $f(x) = kx$ for some real number k. (Claim 7)

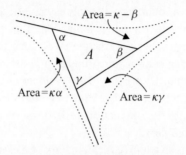

Now, given any triangle with angles α, β, γ, and area A, we extend its sides to three ideal points, Ω, Ω', and Ω'', and let these points be the vertices of a triply asymptotic triangle. (Use Claim 5 to draw its sides.) Having embedded our triangle in a triply asymptotic triangle[12], we express the area of the latter in two ways, and obtain the equation

$$t = A + f(\alpha) + f(\beta) + f(\gamma).$$

Since $t = f(\pi)$, we may rewrite this as $\qquad f(\pi) = A + f(\alpha) + f(\beta) + f(\gamma)$.
Because $f(x) = kx$, this becomes $\qquad\qquad k\pi = A + k\alpha + k\beta + k\gamma$.
Or equivalently, $\qquad\qquad\qquad\qquad\qquad A = k(\pi - \alpha - \beta - \gamma)$.

Therefore, A is directly proportional to defect, as claimed. ∎

One consequence of this relationship between area and angle defect is that triangles in imaginary geometry cannot exceed a certain maximum possible area. (Proof: Area is proportional to defect, and defect can never exceed π, so a triangle's area can never exceed πk, where k is the constant of proportionality.) Interestingly, it can be shown that circles do assume arbitrarily large areas in imaginary geometry; consequently, imaginary geometry contains certain circles so large that no triangle can contain them.

Lambert's Sphere of Imaginary Radius

> "From this I should almost conclude that the third hypothesis would occur in the case of an imaginary sphere."
> —Lambert

By 1766, Lambert knew that if his "third hypothesis" were true, then all triangles would exhibit angle defect, and their defects would be proportional to their areas. In the notes to TP 20, we have already seen some of the thoughts that led him to this conclusion. Faced with this result, and its jarring corollary that triangles' areas are bounded by a finite constant, Lambert

---

[12] This construction shows, incidentally, that no triangle in imaginary geometry can have an area that exceeds t. Thus, the statement, "there exist triangles of arbitrarily large area" is equivalent to the parallel postulate.

avoided the tempting trap of believing that he had found a contradiction in his third hypothesis. Rather, he made the cryptic comment quoted above, which seems to suggest that the third hypothesis might describe the geometry of a "sphere of imaginary radius", whatever that might be.

Presumably, he was struck by the close relationship between the proportion that he had just deduced, which we may express in the form

$$A = k(\pi - \alpha - \beta - \gamma), \tag{3}$$

where k is an unknown constant, and α, β, γ are the angles of a triangle, and the classical formula from spherical geometry relating a spherical triangle's angular *excess* to its area

$$A = r^2(\alpha + \beta + \gamma - \pi), \tag{4}$$

where r is the radius of the sphere. To make geometric sense, the radius r must be a positive number. However, if we work naively and algebraically, without worrying about geometric meaning, we might notice that if we let r be the *imaginary* number $\sqrt{k}i = \sqrt{k}\sqrt{-1}$ (for some positive real number k), then formula (4) will be transformed into formula (3).

The preceding algebraic "coincidence" hints tantalizingly at a possible connection between spherical geometry and imaginary geometry (glimpsed appropriately enough, with the help of imaginary numbers!), but the precise nature of such a relationship, if it indeed exists, remains obscure. We shall return to Lambert's vision of a sphere of imaginary radius at the very end of *The Theory of Parallels*, after Lobachevski's development of imaginary trigonometry.

Theory of Parallels 34

We define a horosphere to be the surface generated by revolving a horocycle about one of its axes, which, together with all the remaining axes of the horocycle, will be an axis of the horosphere.

Just as we can produce a sphere by revolving a circle about one of its diameters[1], we produce a *horosphere* by revolving a horocycle about one of its axes. We shall call this axis the horosphere's *axis of rotation*.

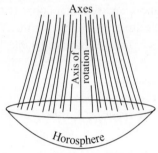

As the generating horocycle revolves about the axis of rotation, each of its axes traces out a trumpet-like cylinder in space. We define all the rays that lie upon these cylinders as the *horosphere's axes* (i.e. its "other" axes, besides its axis of revolution). Clearly, every axis of the horosphere lies on some line in space, parallel to the horosphere's axis of rotation; conversely, any line in space that is parallel to the axis of rotation contains one of the horosphere's axes. Thus, there is a natural correspondence between a horosphere's axes and the pencil of lines parallel to its axis of rotation.

Because of this correspondence, it is tempting to abuse the term "axis"; we will generally use it in its strict sense, to refer to the ray whose endpoint lies on the relevant horocycle or horosphere, but sometimes we will yield to the temptation and use it in a looser sense, to refer to the line containing that ray. The context will always make it clear which meaning is meant.

In TP 34, Lobachevski introduces the horosphere and proves the remarkable fact that its intrinsic geometry is *Euclidean*. His proof contains many sticky details, its basic idea, which follows, is simple. Just as an ordinary sphere has its own intrinsic "lines" (great circles), so a horosphere has its own "lines" (horocycles). Just as spherical geometry concerns points, "lines", and circles on a sphere, so "horospherical geometry" concerns points, "lines", and circles on a horosphere. Lobachevski will show that the points, lines, and circles of horospherical geometry obey all five of Euclid's postulates – including the parallel postulate - and therefore obey every theorem of Euclidean geometry. This is the sense in which the horosphere's geometry is intrinsically Euclidean.

The bulk of TP 34 is devoted to showing that Euclid's *first* postulate (through any two points there is a unique line) holds on the horosphere. As we shall see, we have already done (in TP 28) most of the work that is needed to secure the parallel postulate. Before undertaking a detailed

---

[1] In fact, this is essentially how Euclid defines a sphere in Book XI of the *Elements*.

examination of Lobachevski's proof, we shall establish some preliminary propositions, which will help us to gain a better feel for the horosphere.

Preliminary Propositions

We shall often refer to a horosphere by its axis of rotation, as in, *the horosphere with axis of rotation AA'*. To justify this convention, we must prove that each ray AA' determines a horosphere unambiguously. That is, we must prove that ray AA' is the axis of rotation for only one horosphere.

Claim 1. A horosphere is uniquely determined by its axis of rotation.

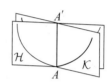

Proof. Let AA' be a ray. By definition, any horosphere with axis of rotation AA' is the result of revolving a horocycle, one of whose axes is AA', about AA'. We must show that all such horocycles generate the same horosphere.

To this end, let $\mathcal{H}$ and $\mathcal{K}$ be any two such horocycles. Because each has AA' as an axis, each lies in a plane containing AA'. That is, the planes upon which $\mathcal{H}$ and $\mathcal{K}$ lie intersect at line AA'. Consequently, we may rotate either plane about AA' to bring it into coincidence with the other. Because $\mathcal{H}$ and $\mathcal{K}$ are symmetric about AA' (TP 31 Notes, Claim 4), and congruent to one another (TP 31 Notes, Claim 5), this rotation will also bring $\mathcal{H}$ and $\mathcal{K}$ into coincidence. Since each horocycle lies in the other's orbit, they will trace out the same horosphere, which we shall call *the* horosphere with axis of rotation AA'. ∎

Those horocycles that lie on the surface of a horosphere will be of particular interest to us. In the geometry of the horosphere, they play the roles of straight lines, as great circles do in spherical geometry. To avoid frequent repetition of the awkward phrase, "horocycle lying on the surface of the horosphere," I shall call such horocycles *surface horocycles*.

Claim 2. Given any point on a horosphere, a unique surface horocycle passes through it and the endpoint of the axis of rotation. (Moreover, the axes of this surface horocycle are axes of the horosphere.)

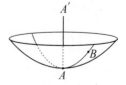

Proof. Let B be an arbitrary point on the horosphere with axis of rotation AA'.

The existence of a surface horocycle through A and B is clear: since B lies on the horosphere, it must have been "hit" by the revolving horocycle (which always passes through A) that traced the horosphere out. Moreover, the axes of such a surface horocycle are clearly axes of the horosphere, by definition of the latter.

Uniqueness is also simple. The plane containing a surface horocycle through A and B must contain ray AA' and point B. Consequently, the only plane that can contain a surface horocycle through A and B is plane $AA'B$. We know that there is a unique horocycle in plane $AA'B$ passing through B, whose "center" is the pencil of parallels that includes AA' (TP 31 Notes, Claim 9). Hence, there is a unique surface horocycle through A and B, as claimed. ∎

Claim 3. The intersection of a horosphere and any plane containing its axis of rotation is a surface horocycle, whose axes are axes of the horosphere.

Proof. Since a horosphere is a surface of revolution, its intersections with any two planes containing its axis of rotation will be congruent: we may bring these intersections into coincidence with one another via a rotation about the axis. Since the horocycle that generates the horosphere obviously arises as such an intersection, it follows that all such intersections must be horocycles. The axes of these horocycles are axes of the horosphere, by definition. ∎

Claim 4. If the axis of rotation of a horosphere also is the axis of some horocycle,[2] then the horocycle lies on the surface of the horosphere. (i.e. it is a surface horocycle.)

Proof. Let AA' be the horosphere's axis of rotation, $\mathcal{H}$ the given horocycle, and T the plane in which $\mathcal{H}$ lies.

Thus, $\mathcal{H}$ is the unique horocycle in plane T which passes through A, and whose "center" is the pencil of parallels containing AA'. Since the intersection of T and the horosphere is a surface horocycle (Claim 3), which obviously also lies in plane T, passes through A, and has AA' for an axis, this surface horocycle is identical to $\mathcal{H}$, since $\mathcal{H}$ is the unique horocycle satisfying these conditions. Thus, the given horocycle $\mathcal{H}$ is a surface horocycle, as claimed. ∎

Recall Gauss' definition of corresponding points with respect to a pencil of parallels: A and B are said to be *corresponding points* with respect to a particular pencil of parallel lines if the rays of the pencil that emanate from A and B are equally inclined to the line segment AB. The definition of corresponding points remains the same in three (or even n) dimensions, and provides the appropriate language for the following alternate characterization of the horosphere:

A horosphere is the surface consisting of all points in space that correspond to the endpoint of a given ray, with respect to the pencil of lines parallel to that ray.

The simple proposition that follows will demonstrate that this alternate characterization, which is essentially how Bolyai defined the horosphere in his *Appendix*, is equivalent to Lobachevski's definition.

Claim 5. Each point B in space lies on the horosphere with axis of rotation AA' if and only if B corresponds to A, with respect to the horosphere's axes.

Proof. ⇒) If B is an arbitrary point on the horosphere, then there is a surface horocycle through B and A (Claim 2), whose axes are axes of the horosphere. Since any two points of a horocycle correspond with respect to its axes (by Bolyai's definition of the horocycle), B and A correspond with respect to the axes of the horosphere.

⇐) If B corresponds to A, then B lies on the unique horocycle in plane BAA' that passes through A and has axis AA'. By Claim 4, this is a surface horocycle. Thus, B lies on this horosphere. ∎

---

[2] Here, we are being strict, thinking of axes not as lines, but *rays*, whose endpoints lie on their associated horocycle/horosphere.

To recapitulate, any ray AA' determines a unique horosphere, which we may characterize in two equivalent ways:

1. The surface of revolution (about AA') generated by any horocycle having ray AA' as an axis.
2. The surface consisting of all points corresponding to A, with respect to the pencil of parallels to AA'.

Towards the Homogeneity of the Horosphere

In the notes to TP 31, we demonstrated that there is nothing special about a horocycle's generating axis. In fact, *any* axis of a horocycle can be viewed as its generator, since horocycles, like circles and straight lines, are *homogeneous* curves: they "look the same" from all of their points. Much of TP 34 is devoted to establishing that the horosphere, like a sphere or plane, is a homogeneous surface. This is important in its own right, and it will help us prove that Euclid's first postulate holds in horospherical geometry. Here too, the key is to show that a horosphere's axis of rotation is no more distinguished than are any of its other axes.

Homogeneity is not a surprising property for a "sphere of infinite radius" to possess, but it does take a fair amount of work to establish it rigorously. Lobachevski will ultimately secure the homogeneity of the horosphere from the following related result: *every point on a horosphere corresponds to every other point upon it, with respect to the horosphere's axes*. Or, as he states it:

> Any chord joining two points of the horosphere will be equally inclined to the axes that pass through its endpoints, regardless of which two points are taken.

It will take several pages to establish this claim. Once we prove it, we will need but a few short steps to deduce the homogeneity of the horosphere. Let us now examine Lobachevski's lengthy proof.

The Long Proof of Homogeneity

> Let A, B, and C be three points on the horosphere, where AA' is the axis of rotation and BB' and CC' are any other axes. The chords AB and AC will be equally inclined toward the axes passing through their endpoints; that is, $\angle A'AB = \angle B'BA$ and $\angle A'AC = \angle C'CA$ (TP 31). The axes BB' and CC' drawn through the endpoints of the third chord BC are, like those of the other chords, parallel and coplanar with one another (TP 25).

Lobachevski establishes the setting on the horosphere with axis of rotation AA'. Upon this surface, he chooses two arbitrary points, B and C. *His goal here, and for the next several pages, is to show that B and C correspond to one another with respect to the horosphere's axes.* To begin, he notes that each of these points corresponds to A. (This follows from Claim 5.)

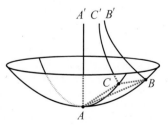

Since AA' and BB' are axes of a horocycle (Claim 2), we have $AA'\|BB'$. Similarly, $AA'\|CC'$. Hence, $BB'\|CC'$ by the transitivity of parallelism. Obviously, BB' and CC' are coplanar, since coplanarity is part of the definition of parallelism.

The proof thus far suggests an infinite prism (of the sort described in TP 28) within the horosphere, built upon triangle $\triangle ABC$; the prism's three infinitely long edges, AA', BB', and CC', are parallel to one another.[3] To simplify subsequent figures in this proof, I shall typically draw the prism alone, as in the figure at right. Although I will be omitting the horosphere in the figures, the reader should remember that A, B, and C are points on a horosphere, whose axis of rotation is AA'.

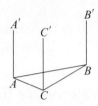

The perpendicular DD' erected from the midpoint D of chord AB in the plane of the two parallels AA', BB' must be parallel to the three axes AA', BB', CC' (TP 31, 25). Similarly, the perpendicular bisector EE' of chord AC in the plane of parallels AA', CC' will be parallel to the three axes AA', BB', CC', as well as the perpendicular bisector DD'.

In the two faces of the prism meeting at AA', Lobachevski constructs DD' and EE', the perpendicular bisectors of AB and AC, respectively.

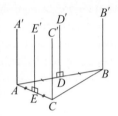

Because AB and AC are chords not only of the horosphere, but also of surface horocycles (Claim 2), their perpendicular bisectors must be parallel to the axes passing through their endpoints (Lobachevski's definition of the horocycle in TP 31). Thus, $DD' \| AA'$, CC' and $EE'\| AA'$, BB'. Since DD' and EE' are both parallel to AA', they are parallel to one another (TP 25).

Hence, the five lines, AA', BB', CC', DD', and EE', are all parallel. Lobachevski's strategy for showing that B and C correspond to one another is to add yet another line to this list: the perpendicular bisector of chord BC, drawn in the remaining face of the prism. Once we establish that this bisector is parallel to BB' and CC', the desired correspondence of B and C will follow easily. However, we shall need to approach the prism's third face in a different manner than the one that we just used for the first two; for here, Claim 2 does not apply, so we do *not* have the luxury of knowing that B and C lie on a surface horocycle.

Denote the angle between the plane of the parallels AA', BB' and the plane in which triangle $\triangle ABC$ lies by $\Pi(a)$, where a may be positive, negative, or zero. If a is positive, draw $DF = a$ in the plane of triangle $\triangle ABC$, into the triangle, perpendicular to chord AB at its midpoint D; if a is negative, draw $DF = a$ outside the triangle on the other side of chord AB; if $a = 0$, let point F coincide with D.

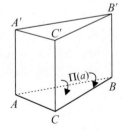

---

[3] If the surface horocycles joining A to B and A to C happen to be identical, then points B and C obviously correspond to one another, since they lie on the same surface horocycle. Since the following lengthy argument, which is designed to prove that correspondence, is unnecessary in this "degenerate case", we shall take the trivial proof for this case for granted, and assume hereafter that A, B, and C do *not* all lie on a single surface horocycle. The degenerate case, moreover, would result in a two-dimensional "degenerate prism." By disposing of this case separately, we may safely assume that our prism is a genuine three-dimensional object.

Lobachevski lets $\Pi(a)$ denote the dihedral angle that the prism's face $AA'BB'$ makes with triangle $\triangle ABC$. This dihedral angle could have any measure between 0 and π. Recalling the graph of the Π-function (at right) that first appeared in the notes to TP 23, we see that

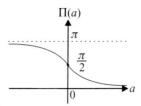

$$a > 0 \Leftrightarrow \text{the dihedral angle } \Pi(a) \text{ is acute.}$$
$$a = 0 \Leftrightarrow \text{the dihedral angle } \Pi(a) \text{ is right.}$$
$$a < 0 \Leftrightarrow \text{the dihedral angle } \Pi(a) \text{ is obtuse.}$$

Whatever the dihedral angle may be, we will have some real number a associated with it. Let DF be a line segment of length $|a|$ that lies upon the perpendicular bisector of AB (in the plane of $\triangle ABC$). For each nonzero value of a, there are two possible locations for F, one on each side of AB. We shall put F on the side of AB that contains the triangle if $a > 0$, and on the other side if $a < 0$.

Although this is not clear now, we shall see later that F turns out to be the circumcenter of $\triangle ABC.$[4]

> All cases give rise to two congruent right triangles, $\triangle AFD$ and $\triangle DFB$, whence $FA = FB$. From F, erect FF' perpendicular to the plane of triangle $\triangle ABC$.

The case in which $F = D$ (when $a = 0$) does not yield a genuine pair of congruent triangles, but it still exhibits the relationship $FA = FB$, which is the important feature across all three cases. Although FF' is defined as the perpendicular erected from plane ABC at point F, the definition of F in terms of a will allow us to prove that FF' is parallel to the five other rays in our picture, a task to which we now turn our attention.

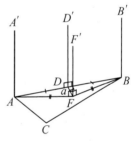

> Because $\angle D'DF = \Pi(a)$ and $DF = a$, FF' must be parallel to DD'; the plane containing these lines is perpendicular to the plane of triangle $\triangle ABC$.

Assuming for the moment that DD' and FF' are coplanar (we shall prove this shortly), our earlier constructions guarantee that they must be parallel to one another. Indeed, the blueprint for their construction was the very picture of parallels developed in TP 16, where Lobachevski introduced the *angle of parallelism*. We have only to verify that $\angle F'FD = \pi/2$ and $\angle FDD' = \Pi(a)$.

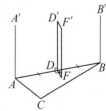

The first equality holds because FF' is perpendicular to plane ABC, and hence to line FD, which lies in that plane. The second holds because $\Pi(a)$, the measure of the dihedral angle between planes $AA'BB'$ and ABC, is defined to be equal to the plane angle

---

[4] Of course, if A, B, and C all happen to all lie on a surface horocycle, then $\triangle ABC$ cannot have a circumcenter in the ordinary sense (TP 31 notes, Claim 7). Such a configuration, however, corresponds to the degenerate case mentioned in the previous footnote, which we have already disposed of; thus, it is irrelevant here.

between any two lines of slope that meet on the dihedral angle's hinge[5]. Since DF and DD' are such lines of slope, it follows that $\angle FDD' = \Pi(a)$.

It remains to verify that DD' and FF' are actually coplanar. In doing so, we will need the following useful lemma from solid neutral geometry.

Claim 6 (Perpendicular plane criterion). Given two planes, if one of them contains a line that is perpendicular to the other, then the two planes are perpendicular.

Proof. Let T and S be the planes. Suppose that WX, a line in T, is perpendicular to plane S, which it intersects at point X. Let XY be the intersection of the two planes. Draw XZ perpendicular to XY in plane S. We must show that the dihedral angle between the planes is $\pi/2$.

Since XZ and XW are lines of slope for this dihedral angle, its measure is equal to $\angle WXZ$. Since WX is perpendicular to plane S, it must be perpendicular to XZ, which lies in S. Thus, $\angle WXZ = \pi/2$. That is, the planes are perpendicular to one another, as claimed. ∎

With the help of this lemma, we return to our promised proof that lines DD' and FF' are coplanar.

Claim 7. DD' and FF' are coplanar.

Proof. Let Γ be the unique plane that contains DF and is perpendicular to plane ABC. We shall show that Γ contains both DD' and FF'.

First, we will show that plane $D'DF$ satisfies the two conditions that uniquely determine Γ.

Since BD is perpendicular to two lines, DF and DD', that lie in $D'DF$, TP 11 implies that BD is perpendicular to $D'DF$. Thus, since BD is contained in ABC, Claim 6 tells us that $D'DF$ is perpendicular to ABC. Since it contains DF as well, $D'DF$ must be Γ. Thus, line DD' lies in plane Γ.

Similarly, plane $FF'D$ contains DF, and it is perpendicular to ABC, since it contains FF', which is perpendicular to ABC, by construction. Hence, $FF'D$ must be Γ. Thus, line FF' lies in Γ. ∎

Having filled this hole, we now know that DD' and FF' are indeed parallel, as claimed. Moreover, the plane containing them is perpendicular to plane ABC.

Moreover, FF' is parallel to EE'; the plane containing them is also perpendicular to the plane of triangle $\triangle ABC$.

Having shown that FF' is parallel to DD', the transitivity of parallelism ushers FF' into our growing list of parallels: we now know that AA', BB', CC', DD', EE', and FF' are all parallel to one another.

Since FF' and EE' are parallel, they are necessarily coplanar. Because it contains FF', plane $EE'FF'$ is, like plane $DD'FF'$, perpendicular to plane ABC, by the perpendicular plane criterion.

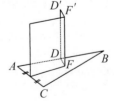

---

[5] See the "Dihedral Digression" in the notes to TP 26.

Next, draw EK perpendicular to EF in the plane containing the parallels EE' and FF'. It will be perpendicular to the plane of triangle $\triangle ABC$ (TP 13), and hence to the line AE lying in this plane. Consequently, AE, being perpendicular to EK and EE', must be perpendicular to FE as well (TP 11). The triangles $\triangle AEF$ and $\triangle CEF$ are congruent, since they each have a right angle, and their corresponding sides about their right angles are equal. Therefore, $FA = FC = FB$.

Here, Lobachevski shows that F is equidistant from the vertices of $\triangle ABC$.

He begins by showing that $EA \perp FE$.

According to TP 13, when two planes are perpendicular to one another, any line that lies in one of them and is perpendicular to their intersection will be perpendicular to the other plane. Thus, having just shown that planes $EE'FF'$ and ABC are perpendicular, we know that EK, which lies in plane $EE'FF'$, is perpendicular to plane ABC. (Incidentally, to verify that EK is distinct from EE', note that $\angle FEE' = \Pi(EF) < \pi/2$, whereas $\angle FEK = \pi/2$.) In particular, $EK \perp EA$, since EA lies in ABC. Thus, $EA \perp EK$ and $EA \perp EE'$ (by definition of EE'), whence TP 11 implies that EA is perpendicular to the plane which contains EK and EE'. Namely, plane $EE'F$. Thus, in particular, $EA \perp FE$, as claimed.

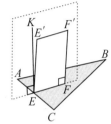

From this, the congruence $\triangle AEF \cong \triangle CEF$ follows, by SAS.

Hence, $FA = FC$.

Combining this with the fact that $FA = FB$ (which Lobachevski noted earlier in the proof), we conclude that $FA = FB = FC$, as claimed. That is, F is the circumcenter of $\triangle ABC$.

In isosceles triangle $\triangle BFC$, a perpendicular dropped from vertex F to the base BC will fall upon its midpoint G.

Let G be the foot of the perpendicular. To see that G is the midpoint of BC, simply note that $\triangle CGF \cong \triangle BGF$ (by RASS— see the notes to TP 10). Consequently, $CG = BG$. That is, G is the midpoint of BC, as claimed.

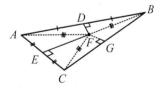

The plane containing FG and FF' will be perpendicular to the plane of triangle $\triangle ABC$, and will cut the plane containing the parallels BB', CC' along a line that is parallel to them, GG'. (TP 25).

Since plane FGF' contains a line, FF', which is perpendicular to plane ABC, these two planes are perpendicular to one another, by the perpendicular plane criterion (Claim 6). Because planes FGF' and $BB'CC'$ share at least one point, G, their intersection must be a line through that point (TP 25 notes: plane axiom 2). Each plane contains one member of a pair of parallels ($FF'\|BB'$), so their line of intersection, which we shall call GG', will also be parallel to the members of this pair (TP 25 Notes, Claim 1: "Lobachevski's Lemma").

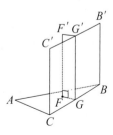

Thus, the seven lines AA', BB', CC', DD', EE', FF', and GG' are all parallel to one another.

Since CG is perpendicular to FG, and thus to GG' as well [TP 13], it follows that $\angle C'CG = \angle B'BG$ (TP 23).

Since planes FGF' and ABC are perpendicular, we may apply TP 13 to them: CG lies in plane ABC and is perpendicular to the line at which these planes meet, so by TP 13, line CG must be perpendicular to plane FGF'. In particular, CG is perpendicular to GG', since it lies in FGF'.

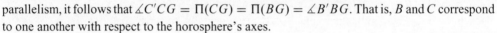

Since $\angle CGG'$ is a right angle and $CC'\|GG'$, we have $\angle C'CG = \Pi(CG)$, by the definition of the angle of parallelism. Similarly, $\angle B'BG = \Pi(BG)$. Finally, because equal lengths (CG and BG) have equal angles of parallelism, it follows that $\angle C'CG = \Pi(CG) = \Pi(BG) = \angle B'BG$. That is, B and C correspond to one another with respect to the horosphere's axes.

Thus, after much work, Lobachevski has finally established that *any two points on a horosphere correspond with respect to its axes*. We may now finally demonstrate the homogeneity of the horosphere.

From this, it follows that any axis of the horosphere may be considered its axis of rotation.

The definition of a horosphere suggests that each horocycle has one particularly distinguished axis, the axis of rotation, which is more fundamental than the others. We have been striving, throughout this long subsection ("The Long Proof of Homogeneity"), to build the tools to debunk this false impression. Finally, we can complete the task.

Claim. The horosphere is a homogeneous surface.

Proof. Consider the horosphere with axis of rotation AA'. We know that AA' determines the horosphere (Claim 1).

Let BB' be an arbitrary axis (other than AA'). We want to prove that BB' also determines the horosphere.

We know that BB' determines some horosphere—the horosphere whose axis of rotation is BB'. We will prove that this second horosphere is identical to the horosphere with which we started (i.e. the horosphere determined by AA').

The axes of the two horospheres lie in the same pencil of parallels, since AA' and BB', their respective axes of rotation, belong to the same pencil of parallels. Consequently, two points in space correspond to one another with respect to the axes of the first horosphere if and only if they correspond to one another with respect to the axes of the second horosphere. Therefore, we will save space below by simply writing that two particular points *correspond*, without bothering to name the axes with respect to which they correspond; such points correspond to one another with respect to the axes of both horospheres simultaneously.

Since B lies on the first horosphere, we know that every point on the first horosphere corresponds to B. Thus, the first horocycle is contained within the second, since the latter is the surface consisting of *all* points that correspond to B (Claim 5). But since all horospheres are congruent to one another,[6] the first horosphere can be contained within the second only if the two are identical. Thus, we have only one horosphere after all, which may be determined either by AA' or by BB'. Accordingly, we may consider either of these axes to be the axis of rotation for this horocycle. Indeed, since BB' was an arbitrary axis of the horosphere, we may consider *any* of its axes to be the axis of rotation.

Since every axis can be considered the axis of rotation, the horosphere "looks the same" from each of its axes, regardless of which point it emanates from. Equivalently, it looks the same from each of its points, and thus is homogeneous, as claimed. ■

Thus, like a sphere or plane, the horosphere offers the same vista to any beholder who stands upon its surface, regardless of the point upon which he stands.

Slices of the Horosphere

We shall refer to any plane containing an axis of a horosphere as a *principal plane*. The intersection of the principal plane with the horosphere is a horocycle; for any other cutting plane, the intersection is a circle.

Here, Lobachevski considers the appearance of all possible cross-sections of a horosphere. The crucial distinction is whether the cutting plane is a *principal plane* (i.e. a plane that contains an axis of the horosphere) or not.

Claim 8. The intersection of a horosphere and a principal plane is a surface horocycle.

Proof. In Claim 3, we showed that the intersection of a horosphere and a plane containing its axis of rotation is a surface horocycle. Since we now know that *every* axis of a horosphere is an axis of rotation, it immediately follows that the intersection of a horosphere and any principal plane is a surface horocycle, as claimed. ■

Claim 9. The intersection of a horosphere and a non-principal plane is a circle.

Proof. Let A, B, C be points in the intersection of a horosphere and a non-principal plane. Let AA', BB', and CC' be the axes emanating from them. Regarding AA' as the horosphere's axis of rotation, we may follow the elaborate constructions of Lobachevski's proof that B and C are corresponding points, and preserve his notation as we proceed. In producing the "prism" upon triangle $\triangle ABC$, we obtain an axis FF' that emanates from the circumcenter of $\triangle ABC$, and is perpendicular to the cutting plane $\triangle ABC$.

When any surface of revolution is cut by a plane perpendicular to its axis of symmetry, the resulting intersection is clearly a circle.[7] This is precisely the situation we have here: since the cutting plane ABC is perpendicular to FF', which we shall think of as the horosphere's axis of rotation, it must cut the horosphere in a circle, as claimed. ■

---

[6] Because all horo*cycles* are congruent and symmetric about each of their axes, the surface produced by rotating any horocycle one of its axes will always "look the same." That is, *all horospheres are congruent.*

[7] In fact, it is a circle in two senses. It is the set of points in plane ABC equidistant (as measured in the plane) from F; it is also the set of points on the horosphere equidistant (as measured along the surface) from the point where FF' intersects the horosphere.

A plane section of an ordinary sphere is always a circle, and when the cutting plane contains one of the sphere's diameters, the intersection will be the largest possible circle on the sphere's surface, a great circle. We have now seen that these phenomena continue to hold on the sphere of infinite radius, the horosphere. A plane section of a horosphere is always a circle; when the cutting plane contains a "diameter", this circle will be as large as the surface of the horosphere allows: it will be a circle of infinite radius—a horocycle. Thus, surface horocycles are to a horosphere as great circles are to an ordinary sphere.

Geometry on the Horosphere

This analogy brings us to the conclusion of TP 34, where Lobachevski extends spherical geometry to spheres of infinite radius—in imaginary space. Just as the "lines" of spherical geometry are the largest circles on the sphere's surface, Lobachevski takes the "lines" of the horosphere to be the its largest circles, the surface horocycles. As exotic and complicated as the prospect of "horospherical geometry" may seem, the reality is that much more startling. Horospherical geometry is Euclidean!

To establish this remarkable fact, we must verify that each of Euclid's five postulates for plane geometry—including the parallel postulate—hold on the horosphere's surface. From this it will follow that every logical consequence of those postulates (i.e. every theorem of Euclidean geometry and trigonometry) also holds on the horosphere.

Any three principal planes that mutually cut one another will meet at angles whose sum is π (TP 28). We shall consider these the angles of a *horospherical triangle*, whose sides are the arcs of the horocycles in which the three principal planes intersect the horosphere. Accordingly, *the relations that hold among the sides and angles of horospherical triangles are the very same that hold for rectilinear triangles in the ordinary geometry.*

Euclid's first two postulates fail spectacularly in spherical geometry. Antipodal points are joined by infinitely many "lines", and line segments can only be extended to the finite length of a great circle. Of course, this is to be expected. Euclid designed his postulates to capture essential aspects of the vast unbounded plane, not of the bounded world of the sphere. Yet this boundedness—the most vital difference between sphere and plane—disappears when the sphere's radius becomes infinite, making the horosphere more plane-like than any sphere of finite radius. Indeed, by interpreting "lines" as "surface horocycles", we may easily show that Euclid's first two postulates hold on the horosphere.

Claim 10. Through any two points on a horosphere, there is a unique line (i.e. a unique surface horocycle).

Proof. Let P and Q be two points on a horosphere. Let PP' and QQ' be the horosphere's axes passing through them. Since any axis of the horosphere may be considered its axis of rotation, we let PP' play this role. Appealing to Claim 2, we conclude that there is a unique surface horocycle joining P and Q, as claimed.   ■

Claim 11. Any line segment (i.e. segment of a surface horocycle) on a horosphere may be extended indefinitely.

Proof. This follows immediately from the fact that horocycles themselves are un-bounded. ■

Euclid's third postulate is the "compass postulate": for any two points *P* and *Q*, there is a circle centered at *P* that passes through *Q*. On an arbitrary surface in space, a circle is defined as the set of points at some fixed distance (measured along the surface) from some fixed point. With this intrinsic definition of a circle, the third postulate automatically holds, in one very trivial sense, on *every* surface: there is always *some* set of points 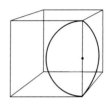 on the surface that satisfy the definition of the required circle, even if this "intrinsic circle" is not a circle in the ordinary sense. For example, the "intrinsic circle" on a cube's surface, centered at the midpoint of one edge and passing through the center of an adjacent face, is not circular in the ordinary sense; it does not lie in a plane.

If this postulate is to have any teeth, it must be more than a mere tautology. Euclid implic-itly makes some assumptions about his circles. In particular, he assumes that they are closed, continuous curves. Naturally, we must verify such "hidden axioms" as well, if we want to be sure that all of Euclidean geometry holds on the horosphere. Fortunately, we need not tease out a list of hidden axioms: instead, we can show that the intrinsic circles of a horosphere (like the intrinsic circles of a finite sphere) are in fact ordinary circles that lie in a particular plane in space; as such, they will possess whatever properties with which Euclid unconsciously endowed them. Consequently, we need not fear that we have forgotten to verify a hidden axiom.

Claim 12. For any points *P* and *Q* on a horosphere, there is a circle on the horosphere centered at *P* passing through *Q*.

Proof. As discussed in the previous paragraph, Euclid's third postulate is, in one sense, trivially true.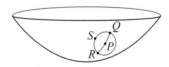

Moreover, the horosphere's intrinsic circles are ordinary planar circles as well, which we may demonstrate as follows. Consider an intrinsic circle with center *P* that passes through *Q*. Extend its radius *PQ* to a diameter *QR*. Let *S* be any other of its points. Since *S* obviously cannot lie on diameter *QR*, which is the unique surface horocycle through *Q* and *R* (Claim 10), the points *Q*, *R*, *S* must be "noncollinear" (i.e., no surface horocycle contains all three of them). It follows that plane *QRS* cannot be a principal plane: if it were, its intersection with the horosphere would be a surface horocycle (Claim 8) containing *Q*, *R*, and *S*. Since *QRS* is a non-principal plane, its intersection with the horosphere is a circle, in both the intrinsic and the ordinary planar senses (Claim 9). Thus, we do have a circle (in *both* senses) centered at *P* that passes through *Q*, as claimed. ■

In general, when two curves in space intersect, we define the measure of the angle between them to be the measure of the angle between their tangent lines. In the TP 26 notes, we have seen an equivalent method for the special case of measuring angles between the "lines" (great circles) of a sphere. Namely, such an angle has the same measure as the dihedral angle between the planes upon which the great circles lie. We can extend this alternate method of measurement to the sphere of infinite radius.

Claim 13. The measure of the angle between two surface horocycles equals the measure of the dihedral angle between the planes in which the surface horocycles lie.

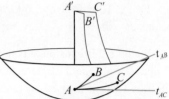

Proof. Consider the angle $\angle BAC$ formed by arcs AB and AC of two surface horocycles. By definition, this angle is measured by the angle between t_{AB} and t_{AC}, the tangents drawn to the horocycles at point A. That is, $\angle BAC = \angle(t_{AB}, t_{AC})$.

The principal plane containing horocycle AB (and its tangent t_{AB}) is $AA'B$, where AA' is the axis of the horosphere passing through A. Similarly, $AA'C$ is the principal plane containing the horocycle AC and its tangent, t_{AC}. To measure the dihedral angle between these principal planes, we must measure the angle between lines of slope that meet at a point on their hinge, AA' (see TP 26 notes: "A Dihedral Digression").

Since horocycles, like circles, have the property that their axes are perpendicular to their tangent lines (TP 31 Notes, Claim 10), we have that $AA' \perp t_{AB}$ and $AA' \perp t_{AC}$. Hence, these tangent lines are lines of slope in the planes $AA'B$ and $AA'C$, respectively. Putting this all together, we have

$$\angle BAC = \angle(t_{AB}, t_{AC}) = \text{(angle between lines of slope)}$$

$$= \text{(dihedral angle between the principal planes)},$$

as was to be shown. ∎

Lobachevski actually *defines* the angle measure between two surface horocycles in terms of the dihedral angle between their principal planes. Lemma 2 assures us that this is a perfectly natural definition, as it agrees with the usual definition in terms of tangent lines.

Regardless of which form of the definition one prefers, it guarantees that the geometry of the horosphere inherits Euclid's fourth postulate ("all right angles are equal") from the geometry of the ambient space in which the horosphere lives, neutral solid geometry.

Claim 14. All right angles are equal on a horosphere.

Proof. Angles on the horosphere are considered "right" if and only if the plane angles between their tangent lines are right. Since all right plane angles are equal by Euclid's fourth postulate (which holds a neutral geometry), all right angles on the horosphere are equal. ∎

Since the first four postulates hold on the horosphere, every theorem of neutral plane geometry holds there. In particular, we know that in neutral geometry, the parallel postulate is equivalent to the statement that the angle sum of all triangles is π. Hence, if we can prove that all horospherical triangles have angle sum π, we will know that the parallel postulate, and hence, *every* theorem of Euclidean plane geometry and trigonometry, holds on the horosphere.

Claim 15. The angle sum of every horospherical triangle is π.

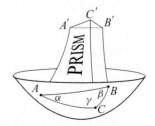

Proof. Let $\triangle ABC$ be a horospherical triangle. Let AA', BB', CC' be the axes emanating from its vertices.

Let α, β, and γ be the measures of angles $\angle A$, $\angle B$, and $\angle C$, respectively. By Claim 13, these angles have the same measure as certain dihedral angles:

α equals the measure of the dihedral angle between planes BAA' and CAA'.

β equals the measure of the dihedral angle between planes ABB' and CBB'.

γ equals the measure of the dihedral angle between planes ACC' and BCC'.

Hence, $\alpha + \beta + \gamma$, the sum of the three angles in the horospherical triangle, equals the sum of the three dihedral angles in the "prism" whose edges are the parallel lines, AA', BB', and CC'. By TP 28, this sum is π. Thus, every horospherical triangle has angle sum π, as claimed. ∎

The horosphere thus emerges as an unexpected Euclidean oasis in the midst of imaginary space. Not surprisingly, Lobachevski adopts this strangely familiar terrain as a base camp from which to conduct further explorations of imaginary geometry. In particular, he will use the horosphere in the remaining propositions to develop imaginary trigonometry.

Taming Wild Geometries

> "In this spirit we have sought, to the extent of our ability, to convince ourselves of the results of Lobachevski's doctrine; then, following the tradition of scientific research, we have tried to find a real substrate for this doctrine, rather than admit the necessity for a new order of entities and concepts."
>
> —Eugenio Beltrami[8]

Imagine the following hypothetical situation. Mathematicians study a system of geometric axioms, contrary to Euclid's own. A series of bizarre theorems—logical consequences of these axioms—emerges: all lines in the plane intersect one another; the area of the entire plane is finite; and the angle sum of every triangle in the plane exceeds π. In the face of such uncomfortable strangeness, one mathematician (a relative of Saccheri?) discovers a spurious "contradiction" and dismisses the novel geometry as a delusion born of logically inconsistent axioms. However, this alleged refutation is soon exposed as wishful thinking, and the irritating geometry remains intact. Years later, another mathematician discovers that the counterintuitive results of this infidel geometry are not so disturbing after all; rather, they describe the intrinsic geometry of a surface in Euclidean space—the sphere (interpreting great circles as lines). The offending system, thus provided with "a real substrate", loses its alien quality. It is accommodated within the larger context of Euclidean space, whose geometry remains *the* geometry of space, and everyone lives happily ever after.

Lobachevski's TP 34 presents a parallel version of this fairy tale. Here, we must imagine the inhabitants of another planet, who are taught from a young age that imaginary geometry (which they simply call *geometry*) is the only possible geometry. Mathematicians in this universe are led to study a system of axioms that imply a host of counterintuitive results. For example, contrary to experience, one can prove that parallel lines in this bizarre geometry do *not* draw closer to one another in the direction of their parallelism. Indeed, parallelism does not even have a direction! After a period of great confusion, an explanation is discovered: these axioms describe the intrinsic geometry of a surface that lies in the traditional space of our fathers and our fathers' fathers.

---

[8] Beltrami, p. 7.

Hence, this surface ultimately derives its strange (Euclidean) intrinsic geometry from the way that it curves within ordinary space. The Euclidean foe having been thus subdued and assimilated into a familiar picture, everyone lives happily ever after in insular bliss.

A civilization that studied and passed imaginary geometry down through the millennia as its sole geometric tradition would thus be able to explain Euclidean geometry away, and continue to maintain that imaginary geometry is the only true geometry of space.

But what of our own civilization? Can we save our Euclidean traditions from the non-Euclidean heresy by taming Lobachevski's geometry—giving it a concrete interpretation as the intrinsic geometry of a surface in Euclidean space?

In 1868, Eugenio Beltrami came very close.[9] He proved that the intrinsic 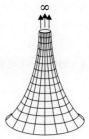 geometry of any surface of constant negative curvature in Euclidean space is a faithful model of a portion of Lobachevski's imaginary plane. The so-called *pseudosphere* (illustrated at right), which resembles an infinitely long trumpet, is an example of such a surface. Given two points on this (or any) surface, the shortest curve on the surface that joins them is called a *geodesic*; in the context of the surface's intrinsic geometry, such a geodesic is defined to be the "line" through those points.[10]

Upon the surface of the pseudosphere, we may fence off an area in which one can safely "play" imaginary geometry, as follows: let one of the lines running down the horn (starting from the rim, and going to infinity) represent a fence. Let the rim itself be a second fence. Imagine a race of tiny two-dimensional creatures who live on the surface, but are unable to burrow under it, fly away from it, or pass over its two fences. These creatures, if they were geometrically inclined, would find that the geometry of their world (whose lines are, naturally enough, its geodesics) is exactly like the geometry of the imaginary plane, with the obvious exceptions imposed by the fences: when drawing lines, for example, the creatures could not extend their line segments past the fences. Consequently, Euclid's second postulate fails on the pseudosphere.[11]

Beltrami's success in finding a "real substrate" for non-Euclidean geometry was therefore only partial. Nevertheless, this partial success was a crucial step in convincing mathematicians that imaginary geometry did somehow partake of "reality", after all.[12] This was but one of Beltrami's many achievements in imaginary geometry, where he played a vital role both as mathematician

---

[9] An English translation of Beltrami's paper is in Stillwell (pp. 7–34).

[10] The "lines" of the sphere and horosphere (great circles and surface horocycles, respectively) are the geodesics of these surfaces.

[11] The first fence wards off topological difficulties. With this fence in place, the pseudosphere is topologically equivalent to a portion of plane; without the fence, it is topologically equivalent to a cylinder. Beltrami circumvented this problem by working with the pseudosphere's *universal cover*—a sort of abstract tissue wrapped infinitely many times about the surface—rather than working with the pseudosphere itself. This allows him to extend line segments past the first fence, but the fence at the rim remains an intractable problem.

[12] Actually, this step should have occurred decades earlier. In 1840, Ferdinand Minding published a study of surfaces of constant negative curvature in Crelle's Journal. In this paper, Minding derived the trigonometric formulae for geodesic triangles on such surfaces (Minding, pp. 323–7). These are identical to the trigonometric formulae for imaginary geometry, which Lobachevski had published three years earlier *in the same journal* (Lobachevskii, "Géométrie Imaginaire"), but no one seems to have connected these two pieces at the time. Beltrami, however, does make explicit reference to Minding in the course of his own work (Beltrami, p. 18).

and as historian—it was he who rediscovered and brought attention to the long-forgotten work of his countryman, Gerolamo Saccheri.

Besides his surfaces of constant negative curvature, Beltrami developed still other models of imaginary geometry of a somewhat more radical nature, which reside not in Euclidean space, but rather in the Euclidean *plane*. Naturally, one must pay a price for the drastic compression that is required to force the entire imaginary plane into the Euclidean plane (and sometimes into a finite portion of Euclidean plane!). Namely, distances can no longer be measured in the ordinary way: one must use a special "ruler" to extract quantitative information from them. Beltrami rarely gets the credit that he deserves for these models, which are most commonly called the "Poincaré disc model", the "Poincaré half-plane model", and "the Klein disc model", after Henri Poincaré and Felix Klein.

In 1901, Hilbert proved that *no* smooth surface in Euclidean space admits an intrinsic geometry that models the *entire* imaginary plane.[13] Consequently, the models of Beltrami are the best possible: if one insists upon embedding Lobachevski's plane geometry into a familiar Euclidean space, then one must be content with either a partial embedding or a warped method of measurement. Beltrami's partial success was in fact the best possible. Although the denizens of a non-Euclidean world can tame Euclidean geometry, the reverse is not quite true. Imaginary geometry refuses to be entirely domesticated.

The Enigma of F.L. Wachter

Early in 1817, a young mathematician named Friedrich Ludwig Wachter published a paper that purported to prove the parallel postulate. The crux of his flawed proof was an attempt to establish that every tetrahedron in space has a circumsphere (cf. Farkas Bolyai's proof in the TP 29 Notes that the parallel postulate would hold if every triangle in the plane had a circumcircle.)

In December 1816, Wachter wrote to Gauss, his former teacher, with whom he had recently discussed "anti-Euclidean geometry". In his letter, Wachter claims that the surface towards which a sphere of increasing radius tends would support Euclidean geometry, even if the parallel postulate were false. Wachter offers no hint of a proof, and his words are far from clear, but they can be read as a remarkable, if hazy, anticipation of the horosphere. Opinions on the value that we should attribute to Wachter's work vary widely.

According to Harold Wolfe,

> *Wachter lived only twenty-five years. His brief investigations held much promise and exhibited keen insight. Had he lived a few years longer he might have become the discoverer of Non-Euclidean Geometry. As it was, his influence was probably considerable. Just at the time when he and Gauss were discussing what they called* Anti-Euclidean Geometry, *the latter began to show signs of a change of viewpoint.*[14]

On the other hand, Jeremy Gray writes,

> *The only hint we have that he* [Gauss] *explored the non-Euclidean three-dimensional case is the remark by Wachter, but what Wachter said was not encouraging: "Now the*

---

[13] Hilbert, pp. 191–199.

[14] Wolfe, p. 56.

inconvenience arises that the parts of this surface are merely symmetrical, not, as in the plane, congruent; or, that the radius on one side is infinite and on the other imaginary" and more of the same. This is a long way from saying, what enthusiasts for Gauss's grasp of non-Euclidean geometry suggest, that this is the Lobachevskian horosphere, a surface in non-Euclidean three-dimensional space on which the induced geometry is Euclidean.[15]

Whatever Wachter may have known or intuited, he had no time to develop his ideas. On the evening of April 3, 1817, he left his house to go for a walk, and never returned. Kurt Biermann has recently argued (based on letters that Wachter's father wrote to Gauss) that the combined blows of a failed love affair and Gauss' judgment on his attempted proof of the parallel postulate may have driven Wachter to despair and suicide.[16]

---

[15] Dunnington, p. 466.
[16] Biermann, pp. 41–43.

Theory of Parallels 35

Lobachevski now begins to develop imaginary trigonometry. As promised at the conclusion of TP 22, he will derive the formulae of imaginary trigonometry both in the plane and on the sphere. This project, the culmination of *The Theory of Parallels*, stretches out over three lengthy propositions. TP 35, the first of the three, is the most difficult proposition in the entire work and offers the most virtuosic display of Lobachevski's genius. Within the pages of this proposition, Lobachevski establishes links between triangles in the plane, on the sphere, and on the horosphere. He completely elucidates the structure of imaginary spherical trigonometry and lays the foundations for imaginary plane trigonometry. The conclusion that he reaches in the former setting is striking – imaginary spherical trigonometry is *identical* to Euclidean spherical trigonometry. That is, Lobachevski proves that the entire subject of spherical trigonometry is part of neutral geometry.

As it is easy to get lost in the intricacies of Lobachevski's arguments in this proposition, I have attempted to break it into smaller, more easily digestible pieces.

Building a Prism/Finding its Dihedral Angles

In what follows, we shall use an accented letter, *e.g.* x', to denote the length of a line segment when its relation to the segment which is denoted by the same, but unaccented, letter is described by the equation $\Pi(x) + \Pi(x') = \pi/2$.

Lobachevski introduces simple notation with complicated verbiage. It simply means that he shall denote the complement of angle $\Pi(x)$ by $\Pi(x')$, and vice-versa.

Let $\triangle ABC$ be a rectilinear right triangle, where the hypotenuse is $AB = c$, the other sides are $AC = b$, $BC = a$, and the angles opposite them are $\angle BAC = \Pi(\alpha)$, $\angle ABC = \Pi(\beta)$. At point A, erect the line AA', perpendicular to the plane of triangle $\triangle ABC$; from B and C, draw BB' and CC' parallel to AA'.

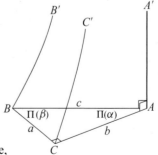

Atop $\triangle ABC$, an arbitrary rectilinear right triangle, Lobachevski constructs the by-now familiar prism. I shall call AA' the *backbone* of the prism to emphasize its special status in the prism's design. Erecting it from vertex A rather than from B was a purely arbitrary choice, as there is nothing to distinguish these two vertices from each other.   Lobachevski addresses this asymmetry later in the proposition by building a second prism, whose backbone emanates from B.

We shall prove an important fact about this prism that Lobachevski uses, but does not prove himself.

Claim 1. $\angle BCC'$ is a right angle.

Proof. Since the backbone AA' is perpendicular to ABC by definition, the perpendicular plane criterion (TP 34 Notes, Claim 6) tells us that every plane containing AA' is perpendicular to ABC. In particular, we know that plane $AA'CC'$ is perpendicular to ABC. We shall now use TP 13 on this pair of perpendicular planes.

Since line BC lies in one of them (ABC) and is perpendicular to the hinge that joins them (AC), TP 13 implies that BC must be perpendicular to the other plane in the pair ($AA'CC'$). Thus, BC is perpendicular to CC', *a fortiori*. That is, $\angle BCC'$ is a right angle, as claimed.   ■

Finally, we note that because Lobachevski has chosen to denote the acute angles of $\triangle ABC$ by $\Pi(\alpha)$ and $\Pi(\beta)$ (rather than α and β), their complements will be $\Pi(\alpha')$ and $\Pi(\beta')$, respectively.

> The planes in which these parallels lie meet one another at the following dihedral angles: $\Pi(\alpha)$ at AA', a right angle at CC' (TP 11 & 13), and therefore, $\Pi(\alpha')$ at BB' (TP 28).

We shall confirm that these dihedral angles are correct:

Claim 2. The measures of the prism's dihedral angles at AA', CC', BB' are $\Pi(\alpha)$, $\pi/2$, and $\Pi\alpha')$, respectively.

Proof. Line AA' is the hinge between planes $AA'BB'$ and $AA'CC'$. Lines AC and AB are lines of slope (see TP 26 Notes: "A Dihedral Digression"), so the dihedral angle at AA' is $\angle BAC = \Pi(\alpha)$, as claimed.

Since $BC \perp AA'CC'$ (see the proof of Claim 1), every plane that contains BC is perpendicular to $AA'CC'$ (by the perpendicular plane criterion). In particular, plane $BB'CC'$ is perpendicular to $AA'CC'$. In other words, the dihedral angle at CC' is $\pi/2$, as claimed.

Finally, since the prism theorem (TP 28) tells us that the three dihedral angles sum to π, the dihedral angle at BB' must be $\pi - [(\pi/2) - \Pi(\alpha)] = \pi/2 - \Pi(\alpha) = \Pi(\alpha')$, as claimed.   ■

Right Triangle Transformation #1: From Rectilinear to Spherical

In this section, Lobachevski shows that each rectilinear right triangle gives rise to a spherical right triangle, whose sides and angles are determined by the sides and angles of the rectilinear triangle from whence it came.

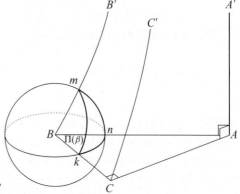

> The points at which the lines BB', BA, BC intersect a sphere centered at B determine a spherical triangle $\triangle mnk$, whose sides are
>
> $$mn = \Pi(c), kn = \Pi(\beta), mk = \Pi(a),$$
>
> and whose opposite angles are, respectively, $\Pi(b)$, $\Pi(\alpha')$, $\pi/2$.

Implicitly, Lobachevski takes the sphere's radius to be the unique value that endows every great circle arc with a length equal to the measure of the angle that it subtends at the sphere's center.[1] Since we will be working with spheres of this particular size throughout this proposition, we shall refer them as *simple spheres*.

Since Lobachevski's sphere is *simple*, it is easy to verify the side lengths of spherical triangle $\triangle mnk$.

$$mn = \angle B'BA = \Pi(c); kn = \angle ABC = \Pi(\beta); \quad \text{and} \quad mk = \angle B'BC = \Pi(a).$$

Next, we verify the angles in spherical triangle $\triangle mnk$.

The angle at vertex k (i.e. the angle opposite side mn) is measured by the dihedral angle between the planes containing its arms, kn and km (see TP 26 notes, Claim 2). This is, of course, the dihedral angle at BC, between planes ABC and $BB'CC'$. Since CC' and CA are lines of slope for it, the measure of this dihedral angle is $\angle C'CA = \Pi(b)$. Thus, the angle of the spherical triangle at vertex k is $\Pi(b)$, as claimed.

Similarly, the angle at m (i.e. the angle opposite side kn) is equal to the dihedral angle of the prism at BB'. This dihedral angle is $\Pi(\alpha')$, as was shown above.

Finally, the angle at n (i.e. the angle opposite side mk) is equal to the dihedral angle between planes $AA'BB'$ and ABC, which is $\pi/2$: these are perpendicular planes by the perpendicular plane criterion, since the former plane contains a line (AA') that is perpendicular to the latter plane.

Bracket Notation for Right Triangles

Suppose we have a right triangle whose legs are a and b, whose hypotenuse is c, and whose respective opposite angles are $\Pi(\alpha)$, $\Pi(\beta)$, and $\pi/2$. If we look down at the plane in which the triangle lies, and read its side lengths, beginning with the hypotenuse, and proceeding counterclockwise around the triangle, we will say either "*c-a-b*" or "*c-b-a*", depending upon the triangle's orientation. A "*c-a-b* triangle" and a "*c-b-a* triangle" are, of course, congruent by the SSS criterion, but they are not *directly* congruent: before sliding them into coincidence, one would need to flip one of the triangles over.

In much of what follows, the orientations of triangles will be important, so we shall introduce notation that indicates not only the sides and angles of a right triangle, but its orientation as well. Specifically,

The "bracket notation" $[a, b, c; \Pi(\alpha), \Pi(\beta)]$ denotes a right rectilinear triangle, whose legs are a, b, whose hypotenuse is c, whose opposite angles are $\Pi(\alpha)$, $\Pi(\beta)$, and $\pi/2$, respectively, and whose sides appear (from above) in the counterclockwise order a-b-c. (Note that the hypotenuse is written last among the sides.)

When the right triangle lies upon a simple sphere, we shall use braces instead of brackets, and we shall order the sides (still cyclic and counterclockwise, with the hypotenuse last) as they appear from the center of the sphere upon which the triangle lies.

---

[1] Here is an intuitive justification for the existence of such a radius. In any circle of radius r, it is clear that a central angle θ cuts off an arc whose length, s, must be proportional to θ. That is, $s = f(r)\theta$, for some function f. (In Euclidean geometry, we know that $f(r) = r$.) Since the circle's circumference vanishes as its radius goes to 0, and becomes arbitrarily large as its radius goes to ∞, it follows that $f(r) \to 0$ as $r \to 0$, and $f(r) \to \infty$ as $r \to \infty$. Consequently, if we make the natural assumption that f varies continuously as a function of r, then we may conclude that there is some value of r for which $f(r) = 1$. Hence, in a circle with this radius, $s = \theta$. By taking this radius as the radius for our sphere, we obtain the desired property that great circle arcs have lengths equal to the central angles they subtend.

Thus, the existence of a rectilinear triangle with sides a, b, c and opposite angles $\Pi(\alpha)$, $\Pi(\beta)$, $\pi/2$ implies the existence of a spherical triangle with sides $\Pi(c)$, $\Pi(\beta)$, $\Pi(a)$ and opposite angles $\Pi(b)$, $\Pi(\alpha')$, $\pi/2$.

Translating this statement into bracket notation, we can summarize Lobachevski's first triangle transformation as follows:

$$[a, b, c; \Pi(\alpha), \Pi(\beta)]$$
$$\Rightarrow \{\Pi(c), \Pi(\beta), \Pi(a); \Pi(b), \Pi(\alpha')\}.$$

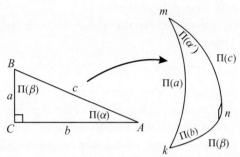

We can rewrite this transformation in an equivalent form, which is easier to use when we wish to apply it to a triangle "in the wild."

$$\textbf{(R} \rightarrow \textbf{S)} \qquad [R_1, R_2, R_3; R_4, R_5] \rightarrow \{\Pi(R_3), R_5, \Pi(R_1); \Pi(R_2), \pi/2 - R_4\}.$$

For the sake of brevity, we shall refer to this transformation of a rectilinear right triangle to a spherical right triangle as "the $R \rightarrow S$ transformation." We shall derive several other such transformation rules shortly.

Once we have secured a few such rules, we shall combine them with blithe algebraic abandon (see Claim 4, below) and derive a further transformation that can look mysterious if one has forgotten the geometric subtleties underlying the formal transformation rules. Lest this occur, I shall recapitulate the geometric aspect of the $R \rightarrow S$ transformation before moving on.

The $R \rightarrow S$ transformation turns a rectilinear right triangle into a spherical right triangle as follows. The given rectilinear triangle becomes the base for an infinite prism, while one of its vertices becomes the center of a simple sphere. Three edges of the prism meet at this vertex; the points where they (or their extensions) pierce the sphere become the vertices of the resulting spherical triangle. More specifically, if we express the rectilinear triangle in bracket notation, then the $R \rightarrow S$ transformation requires us to erect the prism's backbone at the vertex lying opposite the side that occupies the bracket's first slot. (Similarly, the sphere will be centered at the vertex lying opposite the side that occupies the bracket's second slot.) This particular detail – a manifestation of the asymmetry in the prism construction - is easy to overlook, but it will have an important consequence in Claim 4.

Right Triangle Transformation #2: From Spherical to Rectilinear

Conversely, the existence of such a spherical triangle implies the existence of such a rectilinear triangle.

Lobachevski asserts that the $R \rightarrow S$ transformation is invertible, but leaves the details to his reader. I shall carry these out, showing that a right spherical triangle $\triangle mnk$ $\{\Pi(c), \Pi(\beta),$ $\Pi(a); \Pi(b), \Pi(\alpha')\}$ whose legs are both less than $\pi/2$ in length, implies the existence of a right rectilinear triangle $\triangle ABC$ $[a, b, c; \Pi(\alpha), \Pi(\beta)]$. I shall do this in two steps: first, I shall describe how to "build" $\triangle ABC$; second, I shall prove that its bracket notation assumes the required form.

Construction. ($S \rightarrow R$ transformation). Consider a right (simple) spherical triangle with bracket notation $\{\Pi(c), \Pi(\beta), \Pi(a); \Pi(b), \Pi(\alpha')\}$. Assume further that both legs of the triangle have length less than $\pi/2$. Call the triangle $\triangle mnk$, where n denotes the right angle, and k the angle whose measure is $\Pi(b)$.

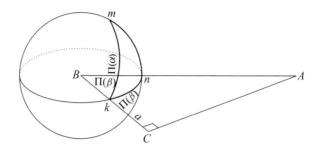

We shall construct $\triangle ABC$ in plane Bnk, which we shall refer to as "the equatorial plane." Let B be the sphere's center. Let C be the unique point on ray Bk such that $BC = a$.[2]

From C, erect a perpendicular in the equatorial plane. It will intersect ray Bn at some point.[3] Call it A. We have now constructed $\triangle ABC$. It remains to prove that it has the required bracket representation. ◆

Note that this transformation has, like the $R \to S$ transformation, a basic asymmetry. Just as the $R \to S$ transformation "favors" one of a given rectilinear triangle's two acute angles (by making it the site at which the prism's backbone is erected), the $S \to R$ transformation favors one of the two legs of the right spherical triangle (by making its plane that in which the rectilinear triangle is constructed.) By favoring the *other* vertex/side in either of these transformations, we would obtain a completely different triangle. This asymmetry is something to keep in mind, as it will return later in the proposition.

We shall now demonstrate that our new transformation works as advertised.

Claim 3. The $S \to R$ transformation just described accomplishes the following:

$$\{\Pi(c), \Pi(\beta), \Pi(a); \Pi(b), \Pi(\alpha')\} \Rightarrow [a, b, c; \Pi(\alpha), \Pi(\beta)].$$

Proof. We must show that in $\triangle ABC$, we have $BC = a$, $AB = c$, $AC = b$, $\angle B = \Pi(\beta)$, and $\angle A = \Pi(\alpha)$.

One side and one angle are easy to verify. **Side BC has length a** (by design). **Angle B has measure $\Pi(\beta)$** (because great circle arc nk subtends a central angle equal to its own length).

To verify the remaining three data requires more work. We begin by constructing a prism upon $\triangle ABC$ as follows. Draw ray Bm, and rename it BB'. In plane $BB'C$, draw $CC' \perp BC$. This line will be parallel to BB', since $\angle C'CB = \pi/2$, $\angle B'BC = \Pi(a)$, and $BC = a$. Finally, draw AA' parallel to BB' and CC', completing the prism. In fact, we can prove that AA' is the

---

[2] The assumption about the triangle's legs implies that the hypotenuse $\Pi(a)$ must also be less than $\pi/2$. (Convince yourself of this by contemplating the figure until it becomes obvious that k must be nearer to m than to the North Pole.) Hence, a must be positive. Were $\Pi(a)$ greater than $\pi/2$, a would be negative, in which case we could not construct C and the argument would founder. This is why we assume that the legs are less than $\pi/2$.

[3] *Proof.* By definition of angle of parallelism, a ray emanating from B will cut the perpendicular if and only if the angle that it makes with BC is less than $\Pi(a)$. In particular, ray Bn and the perpendicular will meet if and only if $\angle nBC < \Pi(a)$. On right triangle $\triangle mnk$, $\Pi(\beta)$ is a leg, while $\Pi(a)$ is the hypotenuse, so we have that $\Pi(\beta) < \Pi(a)$. Since the sphere is simple, $\angle nBC = nk = \Pi(\beta)$. That is, $\angle nBC = \Pi(\beta) < \Pi(a)$, so the ray and the perpendicular meet, as claimed.

backbone of this prism: that is, AA' is perpendicular to plane ABC.[4] Once this is established, we can verify the three remaining data in $\triangle ABC$.

Side *AB* has length *c*.

Proof. Since $AA' \| BB'$, $\angle A'AB = \pi/2$, and $\angle B'BA = \Pi(c)$, this follows from the definition of angle of parallelism.

Side *AC* has length *b*.

Proof. Since $AA' \| CC'$ and $\angle A'AC = \pi/2$, the result will follow if we can show that $\angle C'CA = \Pi(b)$. To this end, we measure the dihedral angle at BC in two different ways. First, because it has the same measure as angle $\angle mkn$ in the spherical triangle (TP 26 Notes, Claim 2), its measure is $\Pi(b)$. Second, because it has lines of slope CC' and CA, its measure is also equal to $\angle C'CA$. Consequently, $\angle C'CA = \Pi(b)$, so $AC = b$, as claimed.

Angle ∠*BAC* has measure Π(α).

Proof. $\angle BAC$ has the same measure as the dihedral angle at AA', since AB and AC are lines of slope. Thus, by the prism theorem (TP 28), $\angle BAC = \pi -$ (dihedral angle at BB') $-$ (dihedral angle at CC').

These last two dihedral angles are easy to determine.

The dihedral angle at BB' is $\Pi(\alpha')$, since it has the same measure as angle $\angle nmk$ in the spherical triangle.

Since $BB'CC'$ contains a line, BC, which is perpendicular to plane $AA'CC'$ (by TP 11 / Euclid XI.4), it follows that $BB'CC' \perp AA'CC'$ (by the perpendicular plane criterion). That is, the dihedral angle at CC' is $\pi/2$.

Hence, $\angle BAC = \pi -$ (dihedral angle at BB') $-$ (dihedral angle at CC')
$$= \pi - \Pi(\alpha') - \pi/2$$
$$= \pi/2 - \Pi(\alpha')$$
$$= \Pi(\alpha), \text{ as claimed.}$$

Thus, we have shown that the parts of $\triangle ABC$ have all the measurements that we expected. That is, $\{\Pi(c), \Pi(\beta), \Pi(a); \Pi(b), \Pi(\alpha')\} \Rightarrow [a, b, c; \Pi(\alpha), \Pi(\beta)]$, which was to be shown. ■

Since not every spherical right triangle that we meet will have its bracket notation already expressed in the form $\{\Pi(c), \Pi(\beta), \Pi(a); \Pi(b), \Pi(\alpha')\}$, it will be useful to rewrite the transformation with completely general symbols in the five slots of the spherical triangle's bracket notation:

$$(S \to R) \qquad \{S_1, S_2, S_3; S_4, S_5\} \Rightarrow [\Pi^{-1}(S_3), \Pi^{-1}(S_4), \Pi^{-1}(S_1); \pi/2 - S_5, S_2].$$

In Praise of R↔S Transformations: Linked Generic Triangles Yield Two Trigonometries

Developing trigonometry in the plane amounts to discovering relations among the parts of a generic right rectilinear triangle. Lobachevski's generic triangle is $\triangle ABC\ [a, b, c; \Pi(\alpha), \Pi(\beta)]$.

---

[4] *Proof.* Spherical angle $\angle mnk = \pi/2$ has the same measure as the dihedral angle between the planes $AA'BB'$ and ABC (TP 26 Notes, Claim 2). That is, $AA'BB' \perp ABC$. Since a line in ABC (namely, BC) is perpendicular to plane $AA'CC'$ (TP 11 / Euclid XI.4), the perpendicular plane criterion implies that $ABC \perp AA'CC'$. Because planes $AA'BB'$ and $AA'CC'$ are both perpendicular to ABC, their line of intersection is also perpendicular to ABC (Euclid XI.19 – a neutral theorem). That is, $AA' \perp ABC$, as claimed.

Every right rectilinear triangle admits bracket notation of this form, since every acute angle is the angle of parallelism for some length (TP 23).

The right spherical triangle, Δmnk $\{\Pi(c), \Pi(\beta), \Pi(a); \Pi(b), \Pi(\alpha')\}$, is almost, but not quite, generic. Because, as noted in the preceding construction, we have implicitly assumed that Δmnk's legs are less than $\pi/2$, Δmnk is actually the generic right spherical triangle *whose legs are less than $\pi/2$ in length.* Nonetheless, this caveat is of minor importance, for once we have found the trigonometric relations for Δmnk, a simple argument will extend them to *all* right spherical triangles. Hence, we are justified in thinking of Δmnk as the generic right spherical triangle, and in thinking of the $R \leftrightarrow S$ transformations as the links between generic right triangles in the plane and on the sphere.

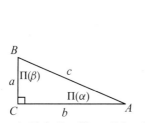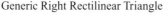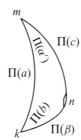

Generic Right Rectilinear Triangle   Generic Right Spherical Triangle
(with legs shorter than $\frac{\pi}{2}$)

This link between the generic triangles ΔABC and Δmnk is extraordinarily powerful.

Because the parts of the two triangles are so closely related, every trigonometric relationship that we discover for spherical right triangles will immediately yield a dual version for rectilinear right triangles, and vice versa. By exploiting this duality, Lobachevski is able to develop trigonometry in the plane and on the sphere simultaneously.

Right Triangle Transformation #3: Reflection (Orientation Reversal)

Reflecting a rectilinear triangle in a line (or reflecting a spherical triangle in a great circle) yields a new triangle, which is congruent to the original, but with reversed orientation. We can capture this geometric operation with a particularly simple bracket transformation rule, which applies both to rectilinear and spherical right triangles. We shall call it the "O-transformation", for orientation reversal.

(O) $[R_1, R_2, R_3; R_4, R_5] \Leftrightarrow [R_2, R_1, R_3; R_5, R_4]$, or $\{S_1, S_2, S_3; S_4, S_5\} \Leftrightarrow \{S_2, S_1, S_3; S_5, S_4\}$.

That is, we swap the first two symbols within the brackets/braces, and swap the last two as well. Geometrically, the two triangles in the O-transformation are mirror images of one another; the existence of one obviously implies the existence of the other.

Right Triangle Transformation #4: Lagniappe

Indeed, the existence of such a spherical triangle also implies the existence of a second rectilinear triangle, with sides a, α', β and opposite angles $\Pi(b'), \Pi(c), \pi/2$.

Hence, we may pass from a, b, c, α, β to b, a, c, β, α, and to a, α', β, b', c, as well.

Lobachevski obtains his fourth transformation by applying the first three successively to a given rectilinear triangle. That is, given a rectilinear triangle, he first converts it into a spherical triangle (using the $R \to S$ transformation), then changes the orientation of the resulting spherical triangle (O-transformation), and finally, turns the result back into a rectilinear triangle ($S \to R$ transformation).

One might guess that the first and third steps would cancel one another out, reducing this overall process to a roundabout application of the O-transformation. In fact, this is not the case at all. Instead, this composite transformation turns $[a, b, c; \Pi(\alpha), \Pi(\beta)]$ into $[a, \alpha', \beta; \Pi(b')$, $\Pi(c)]$, which is clearly *not* a mere reflection of the original triangle. We shall call this process, which changes a rectilinear triangle into a second rectilinear triangle, the "L-transformation." (L is for Lobachevski or Lagniappe.)

Claim 4. (L-transformation) The existence of one rectilinear triangle immediately implies the existence of another:

$$\textbf{(L)} \qquad [a, b, c; \Pi(\alpha), \Pi(\beta)] \Rightarrow [a, \alpha', \beta; \Pi(b'), \Pi(c)].$$

Proof.

$$\begin{aligned}
[a, b, c; \Pi(\alpha), \Pi(\beta)] &\Rightarrow \{\Pi(c), \Pi(\beta), \Pi(a); \Pi(b), \Pi(\alpha')\} &&\text{(by } R \to S) \\
&\Rightarrow \{\Pi(\beta), \Pi(c), \Pi(a); \Pi(\alpha'), \Pi(b)\} &&\text{(by O)} \\
&\Rightarrow [a, \alpha', \beta; \Pi(b'), \Pi(c)]. &&\text{(by } S \to R)
\end{aligned}$$

The L transformation is easy to deduce algebraically, but remains enigmatic geometrically. Why does it not merely change the orientation of the original triangle? To better understand the question, consider the figure at right, which shows all but the last step of the L transformation: we begin with Δ_1, convert it to a spherical triangle, Δ_2, and then apply the O-transformation to the result, yielding Δ_3. To complete the L transformation, we need only convert Δ_3 back into a rectilinear triangle. At first glance, one might assume that this will turn Δ_3 into $\Delta A''BC$, but were this the case, then the L transformation would amount to

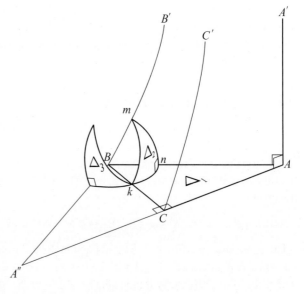

nothing more than the humble O-transformation. What happens in that last step?

The answer lies in the asymmetry of the $R \leftrightarrow S$ transformations. In the last step, *if* the $S \to R$ transformation had "favored" the leg of Δ_3 that lies in plane ABC, then it would indeed turn Δ_3 into $\Delta A''BC$. However, *it favors the other leg*, with the result that we obtain another rectilinear triangle altogether.

Finally, we note in passing that the L transformation is its own inverse.

In Praise of the L-transformation: New Trigonometric Relations from Old

The four transformations that we have seen yield not only new triangles, but also new trigonometric relations. If two triangles are linked by a transformation, then any relation among the parts of one triangle immediately yields a relation among the parts of the other triangle as well. Applied to the $R \leftrightarrow S$ transformations, this duality leads to a simultaneous development of rectilinear and spherical trigonometry. Applied to the L-transformation, this duality leads to a doubling of theorems about rectilinear triangles. To demonstrate this theorem doubling in an easily understood setting, we offer the following simple (albeit insignificant) example.

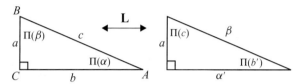

Example. (TP 9) In any rectilinear right triangle, the hypotenuse is longer than either leg.

Applying TP 9 to $\triangle ABC$ yields $a < c$.

Applying the L-transformation to $\triangle ABC$ yields a second triangle, $\triangle XYZ$ $[a, \alpha', \beta; \Pi(b')$, $\Pi(c)]$.

Applying TP 9 to $\triangle XYZ$ yields $a < \beta$.

Re-reading this as a statement about the parts of $\triangle ABC$, we obtain the "bonus" dual theorem.

(TP 9—dualized) *A leg of any rectilinear right triangle is shorter than that length whose angle of parallelism is equal to the acute angle adjacent to that leg.* ◆

The particular dualized theorem in this example is of little consequence, but the process by which we obtained it is not. Because we shall soon use it to obtain results that are more significant, we shall recapitulate the technique, and state an algebraic shortcut for carrying it out.

Given a trigonometric relationship that holds on our generic right rectilinear triangle $\triangle ABC$, we apply the L-transformation to the triangle, interpret the trigonometric relationship on its image $[a, \alpha', \beta; \Pi(b'), \Pi(c)]$, and finally, reinterpret the result as a second relationship on the original triangle.

Formally, we obtain the second relationship by substituting the symbols that occupy corresponding bracket slots of the L-transformation, $[a, b, c; \Pi(\alpha), \Pi(\beta)] \Rightarrow [a, \alpha', \beta; \Pi(b'), \Pi(c)]$. That is, we can dualize any trigonometric relation on $\triangle ABC$ by replacing each b in the relation with an α', each c with β, each $\Pi(\alpha)$ with $\Pi(b')$, and each $\Pi(\beta)$ with $\Pi(c)$.

Right Triangle Transformation #5: From Rectilinear to Horospherical

Lobachevski has already established links between right plane triangles and right spherical triangles. Linking these to right *horospherical* triangles gives him the potential to use the Euclidean trigonometry of horospherical triangles to discover trigonometric relationships that govern triangles (plane and spherical) in imaginary space.

If the horosphere through A with axis AA' cuts BB' and CC' at B'' and C'', its intersections with the planes formed by the parallels produce a horospherical triangle with sides $B''C'' = p$, $C''A = q$, $B''A = r$ and opposite angles $\Pi(\alpha')$, $\Pi(\alpha)$, $\pi/2$.

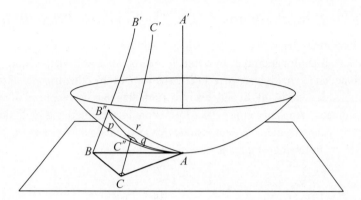

If we think of the plane in which $\triangle ABC$ lies as a vast tabletop, and $\triangle ABC$ itself as a triangle drawn upon it, then we may describe Lobachevski's method for transforming $\triangle ABC$ into a horospherical triangle as follows:

First, place a horosphere on the table; position it so that their point of contact is at the vertex A. Next, draw AA' perpendicular to the table, and then draw rays BB' and CC' parallel to AA'. Finally, slide points B and C up their respective rays, like beads on strings, until they meet the horosphere; let B'' and C'' be the names of the points at which they meet the horosphere. We have thus "projected" the rectilinear triangle $\triangle ABC$ onto the horosphere, producing the horospherical triangle $\triangle AB''C''$.

Since we currently lack the means to describe the side lengths of $\triangle AB''C''$ entirely in terms of the sides and angles of $\triangle ABC$, Lobachevski simply calls them p, q, and r for now, as depicted in the figure above.

In contrast, the angles of the horospherical triangle are easy to determine.

Claim 5. $\angle C''AB'' = \Pi(\alpha)$, $\angle AB''C'' = \Pi(\alpha')$, and $\angle B''C''A = \pi/2$.

Proof. Any angle of any horospherical triangle has the same measure as the dihedral angle between the planes in which its arms lie (TP 34 Notes, Lemma 2). Thus the angle of the horospherical triangle at A, B'', and C'' are measured by the prism's dihedral angles at AA', BB', and CC', respectively, which are $\Pi(\alpha)$, $\Pi(\alpha')$ and $\pi/2$, respectively, by Claim 2. ∎

Although we do not yet know how p, q, and r relate to the measurements of the plane triangle $\triangle ABC$, we do at least know how they relate to one another.

Consequently (TP 34),

$$p = r \sin \Pi(\alpha) \quad \text{and} \quad q = r \cos \Pi(\alpha).$$

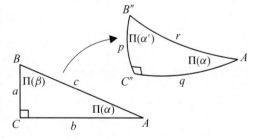

These follow directly from the formulae of Euclidean trigonometry, which hold on the horospherical triangle $\triangle AB''C''$. Writing p and q in terms of r (and $\Pi(\alpha)$, which is also an angle in $\triangle ABC$) effectively reduces our unknowns on the horospherical triangle by two-thirds. This triangle transformation may seem less satisfying than the first four, since we have not succeeded in expressing the sides of the image triangle (on the horosphere) as functions of the original triangle's sides and angles, yet this transformation represents our first step down the winding path that will eventually lead us to the trigonometric formulae that we seek.

The Winding Path: An Overview

The path proceeds roughly as follows. Using the rectilinear-to-horospherical triangle transformation, we shall establish a bizarre-looking equation that expresses a relationship among a motley group of geometric quantities.[5] I call this the "diamond in the rough", for by polishing this equation diligently, we will eventually produce our first trigonometric relationship.

Applying triangle transformations to this polished diamond then yields four more trigonometric relations. Collectively, I call the five relations "the five gems". Remarkably, these will generate *all* of the trigonometric relations that hold in the absence of the parallel postulate, both in the plane and on the sphere. Extracting all of imaginary trigonometry from them does take some work, however. Lobachevski begins this work by producing the formulae of imaginary spherical trigonometry. Astonishingly, these turn out to be identical to those that hold in Euclidean spherical trigonometry! This revelation that the formulae of spherical trigonometry are independent of the parallel postulate marks the end of TP 35.

In TP 36, Lobachevski will finally obtain an explicit formula for the angle of parallelism, which he will then use in TP 37 to extract the formulae of imaginary plane trigonometry from the five gems.

Now that we have some conception of what lies ahead, we begin the journey.

Unfolding the Prism: Mining for the Diamond in the Rough

Along BB', break the connection of the three principal planes, turning them out from one another so that they lie in a single plane. In this plane, the arcs p, q, r unite into an arc of a single horocycle, which passes through A and has axis AA'.

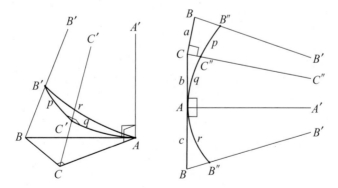

That is, we slit the prism along one of its seams (BB') and then unfold it, as follows: rotate face $BB'AA'$ about AA' until it lies in plane $AA'CC'$, and then rotate face $BB'CC'$ about CC' until it

---

[5] Specifically, it specifies a relationship between various parts of the triangles, and parts of the prism itself, which are involved in the rectilinear-to-horospherical triangle transformation. Namely, one angle of the rectilinear triangle, the hypotenuse of the horospherical triangle, a line segment joining two of the triangles' corresponding vertices, and finally, a horocyclic arc that is concentric to one of the horospherical triangle's legs. This will become clear shortly!

lies in plane $AA'CC'$, as well. We have now unfolded the prism and laid it out in a single plane, as depicted at right. It is easy to see that the three horocyclic arcs fit together smoothly. We prove this now.

Claim 6. When we unfold the prism, the arcs p, q, and r will all lie on a single horocycle.

Proof. Let $\mathcal{H}$ be the unique horocycle of which q is an arc.

In the plane containing the flattened prism, there is a unique horocycle through A with axis AA'. Its uniqueness implies that the horocyclic arcs q and r both lie upon it. In fact, this horocycle must be $\mathcal{H}$, since it contains q. Hence, $\mathcal{H}$ contains both q and r. Similar considerations regarding the unique horocycle through C'' with axis $C''C'$ reveal that $\mathcal{H}$ contains both q and p. Hence, the arcs p, q, and r all lie on the single horocycle $\mathcal{H}$. ∎

Thus, the following lie on one side of AA': arcs p and q; side b of the rectilinear triangle, which is perpendicular to AA' at A; axis CC', which emanates from the endpoint of b, then passes through C'', the join of p and q, and is parallel to AA'; and the axis BB', which emanates from the endpoint of a, then passes through B'', the endpoint of arc p, and is parallel to AA'. On the other side of AA' lie the following: side c, which is perpendicular to AA' at point A, and axis BB', which emanates from the endpoint of c, then passes through B'', the endpoint of arc r, and is parallel to AA'.

This awkward passage simply describes the features of the unfolded prism. It was presumably intended as a verbal substitute for the figure that should have accompanied the text but was in fact consigned (like all the other figures) to a set of plates at the back of the book. This practice was quite common in the 19th century as a means of keeping printing costs within reasonable bounds.

Lobachevski will now introduce some notation for an aspect of the unfolded prism, which will appear shortly as a term in the diamond in the rough.

A New Function: f

The length of the line segment CC'' depends upon b; we shall express this dependence by $CC'' = f(b)$. Accordingly, $BB'' = f(c)$.

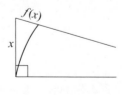

Lobachevski's new function, $f(x)$, is explained by the figure at the right. There are several equivalent ways to describe this function verbally.

Here is one. Draw a tangent line to a point on a horocycle. Upon this tangent, measure out a segment of length x, starting at the point of tangency. The distance from the segment's endpoint to the horocycle is $f(x)$.

Since horocycles are homogeneous, the point from which we draw the tangent will not affect the value of $f(x)$. Since all horocycles are congruent, the particular horocycle upon which we carry out this operation is also irrelevant. Hence, f is well defined.

If we draw the horocyclic arc t that begins at C, has axis CC', and ends at D, its intersection with axis BB', then $BD = f(a)$, so that $BB'' = BD + DB'' = BD + CC''$. That is,

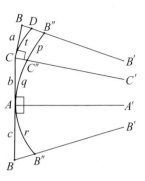

$$f(c) = f(a) + f(b).$$

For the most part, we can read these facts directly from the figure. Note that $DB'' = CC''$ since both of these lengths represent the distance between the concentric horocycles of which p and t are arcs. (See TP 33 Notes, Claim 1.) The relationship $f(c) = f(a) + f(b)$ will prove useful shortly.

The Diamond in the Rough

While following Lobachevski's intricate arguments and marveling at his flights between plane, sphere, and horosphere, it is easy to lose track of the fact that his underlying goal is to develop trigonometric relationships. As I discussed above ("The Winding Path"), these will ultimately derive from the "five gems", a set of trigonometric equations that themselves have a common source in a single "diamond". The diamond, however, will hardly be recognizable as such when we first meet it in the rough.

The trigonometric formulae that we seek must relate the quantities a, b, c, α, and β, which occur (sometimes cloaked in the Π-function) as sides and angles of $\triangle ABC$ and $\triangle mnk$, our representative rectilinear and spherical right triangles. The diamond in the rough exhibits a mixture of these desirable quantities with several unwanted quantities such as t, r, and the f-function, which are only indirectly related to the triangles. Lackluster though it may seem initially, Lobachevski will find a way to scrape off the unwanted bits, and polish what remains into pure sparkling trigonometry.

Moreover, we see that (by TP 33)

$$t = pe^{f(b)} = r \sin \Pi(\alpha)e^{f(b)}.$$

Claim 7. (Diamond in the Rough) $t = r\sin\Pi(\alpha)e^{f(b)}$.[6]

Proof. Applying TP 33 to t and p yields $p = te^{-CC''}$.
Equivalently, $t = pe^{CC''}$.
But $CC'' = f(b)$, so this becomes $t = pe^{f(b)}$.
Thus, since $p = r\sin\Pi(\alpha)$,
(derived just after Claim 5), $t = r\sin\Pi(\alpha)e^{f(b)}$, which was to be shown. ∎

We now have our diamond in the rough. Bit by bit, Lobachevski will polish away its unwanted parts. First, he will scrape away the unwanted horocyclic arcs t and r. Then he will polish off

---

[6] To derive this, we need the formula $s' = se^{-x}$ from TP 33. Recall that a more general form of this equation is $s' = se^{-x/k}$, where k is some positive constant. (Lobachevski picks his unit of length so that $k = 1$.) It is easy to see that if we use this general form of TP 33 in the proof of Claim 7, the diamond in the rough assumes its more general form: $t = r \sin \Pi(\alpha)e^{f(b)/k}$.

the f-function, revealing the diamond at last. The last bit of sediment, the Π-function, can be removed after TP 36, with the help of the explicit formula for $\Pi(x)$ that Lobachevski derives in that proposition.

Seeking the Diamond Within
(Part 1: Scraping Off the Horocyclic Arcs)

The next passage in Lobachevski's text is particularly obtuse. I considered tampering with his words to render them more comprehensible, but to do so would require adding so much more text that the result would cease to be what Lobachevski actually wrote. Thus, I have left it as is, with the promise of an explanation afterwards.

> If we were to erect the perpendicular to triangle $\triangle ABC$'s plane at B, instead of A, then the lines c and r would remain the same, while the arcs q and t would change to t and q, the straight lines a and b would change to b and a, and the angle $\Pi(\alpha)$ would change to $\Pi(\beta)$. From this it follows that
>
> $$q = r \sin \Pi(\beta) e^{f(a)}.$$
>
> Thus, by substituting the value that we previously obtained for q, we find that
>
> $$\cos \Pi(\alpha) = \sin \Pi(\beta) e^{f(a)}.$$

Before worrying about the details of erecting a perpendicular at B rather than A, let us return to our "diamond in the rough": $t = r\sin\Pi(\alpha)e^{f(b)}$.

We obtained this formula by constructing a prism on the right triangle $\triangle ABC$. Because this triangle has no special properties (other than a right angle at C), our decision to erect the prism's backbone at A was a purely arbitrary choice. Had we erected the backbone from B, the resulting formula would have been slightly different. We shall determine the form that it would take in a moment. As a means of doing so, we shall first express our diamond-in-the-rough rhetorically, as though we were 16th century algebraists.

To this end, consider the naked triangle, shorn of all labels. We erect the prism's backbone from one of its two acute vertices (which we shall call *the favored vertex*) and carry out the construction detailed above, thus obtaining a prism, a horospherical triangle, and an "extra" horocyclic arc, concentric to one side of the horospherical triangle and passing through the vertex of the rectilinear triangle's right angle.

The diamond in the rough, expressed rhetorically, would then take the following form:

$$\begin{pmatrix} \text{length of the "extra"} \\ \text{horocyclic arc} \end{pmatrix} = \begin{pmatrix} \text{length of the horospherical} \\ \text{triangle's hypotenuse} \end{pmatrix}$$

$$\times \sin \begin{pmatrix} \text{angle measure at} \\ \text{the favored vertex} \end{pmatrix} e^{f\begin{pmatrix} \text{length of the rectilinear} \\ \text{triangle's leg that touches} \\ \text{the favored vertex} \end{pmatrix}}$$

Having obtained our rhetorical formula, we restore the labels to our triangle, $\triangle ABC$. Now, if we erect the prism upon it so that its backbone emanates from B rather than A, we may apply this rhetorical formula to its parts to obtain another formula, related to our diamond in the rough.

Of course, to carry this out, we must find the quantities on the second prism that occur in the rhetorical formula. This is a simple matter, requiring only the following observation.

Lemma. *Let $\triangle XY\Omega$ and $\triangle X'Y'\Omega'$ be two singly asymptotic right triangles, with right angles at X and X'. If their finite sides XY and $X'Y'$ have the same length, then the horocyclic arcs shown in the figure must also have the same length.*

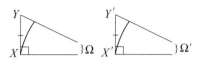

Sketch of Proof. The asymptotic triangles are congruent. Thus, we may bring them into coincidence with one another. Superimposing $\triangle X'Y'\Omega'$ upon $\triangle XY\Omega$ forces the horocyclic arcs to coincide, since there is only one horocycle through X with axis XX'. ∎

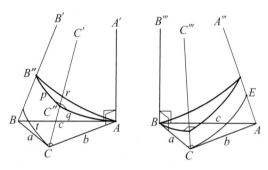

With this lemma in mind, it is easy to discover the lengths of the parts on the second prism that we need for our rhetorical formula; we simply look at the two prisms side by side. For example, the lemma immediately implies that the "extra" horocyclic arc on the second prism (CE) has the same length as the arc AC'' that lies on the first. That is, arc CE has length q. Similarly, the hypotenuse of the horospherical right triangle in the second prism must be r. Finally, we don't need the lemma to tell us that the angle at the favored vertex B is $\Pi(\beta)$, nor that BC, the unique leg of $\triangle ABC$'s that touches B, has length a. Feeding all of this information into the rhetorical formula yields

$$q = r \sin \Pi(\beta) e^{f(a)}.$$

From here, it is a short step to a semi-polished diamond.

Claim 8. (The Diamond, Still Rough, but Semi-Polished) $\cos\Pi(\alpha) = \sin\Pi(\beta)e^{f(a)}$.[7]

Proof. We have just found that $q = r\sin\Pi(\beta)e^{f(a)}$. Moreover, we know that $q = r\cos\Pi(\alpha)$. (We established this just after Claim 5.) Equating the two expressions for q yields the semi-polished diamond:

$$\cos \Pi(\alpha) = \sin \Pi(\beta)e^{f(a)}. \quad ∎$$

This expression is a definite improvement over our original diamond in the rough. The irrelevant horocyclic arcs t and r have been scoured away, bringing us one step closer to a comprehensible trigonometric relation. The next procedure will be to remove the function f.

Seeking the Diamond Within
(Part 2: Polishing Away the f -Function)

Lobachevski has now detached two horocyclic arcs from the emerging diamond. He will remove the function $f(x)$ next. To begin the process, he shows that the related function $\sin\Pi(x)e^{f(x)}$

---

[7] More generally, this will be $\cos\Pi(\alpha) = \sin\Pi(\beta)e^{f(a)/k}$, where k is a positive constant. (See the previous footnote.)

assumes the same value when x is $a, b,$ or c. This unexpected invariant of $\triangle ABC$'s sides will allow us to express $f(x)$ in terms of $\Pi(x)$.

If we change α and β into b' and c, then

$$\sin \Pi(b) = \sin \Pi(c)e^{f(a)}.$$

Multiplying by $e^{f(b)}$ yields

$$\sin \Pi(b)e^{f(b)} = \sin \Pi(c)e^{f(c)}.$$

Consequently, it follows that

$$\sin \Pi(a)e^{f(a)} = \sin \Pi(b)e^{f(b)}.$$

Because the lengths a and b are independent of one another and, moreover, $f(b) = 0$ and $\Pi(b) = \pi/2$ when $b = 0$, it follows that for every a,

$$e^{-f(a)} = \sin \Pi(a).$$

To establish this, we shall finally employ our triangle transformations.

Claim 9. In an arbitrary right rectilinear triangle $\triangle ABC$ $[a, b, c; \Pi(\alpha), \Pi(\beta)]$, the following relationship holds:

$$\sin \Pi(b) = \sin \Pi(c)e^{f(a)}.$$

Proof. Claim 8, we obtained our semi-polished diamond, $\cos \Pi(\alpha) = \sin \Pi(\beta)e^{f(a)}$. We shall use the technique described above, in "The Power of Triangle Transformations." Applying the L-transformation[8] to the semi-polished diamond yields $\cos\Pi(b') = \sin\Pi(c)e^{f(a)}$.

By definition, $\Pi(b')$ is $\pi/2 - \Pi(b)$. Hence, $\cos\Pi(b') = \sin\Pi(b)$, so

$$\sin \Pi(b) = \sin \Pi(c)e^{f(a)}, \text{ as claimed.} \quad \blacksquare$$

We shall now obtain the invariant.

Claim 10. In $\triangle ABC$, the following relation holds:

$$\sin \Pi(a)e^{f(a)} = \sin \Pi(b)e^{f(b)} = \sin \Pi(c)e^{f(c)}.^9$$

Proof. $\sin\Pi(b) = \sin\Pi(c)e^{f(a)}$. (Claim 9)

$\sin\Pi(b)e^{f(b)} = \sin\Pi(c)e^{f(a)}e^{f(b)}$ (Multiplying both sides by $e^{f(b)}$)

$\qquad\qquad = \sin\Pi(c)e^{f(a)+f(b)}$

$\qquad\qquad = \sin\Pi(c)e^{f(c)}$ (since $f(c) = f(a) + f(b)$; see "A New Function: f").

That is, $\sin\Pi(b)e^{f(b)} = \sin\Pi(c)e^{f(c)}$.

Hence, $\sin\Pi(a)e^{f(a)} = \sin\Pi(c)e^{f(c)}$.   (Applying the O-transformation)

---

[8] $[a, b, c; \Pi(\alpha), \Pi(\beta)] \Rightarrow [a, \alpha', \beta; \Pi(b'), \Pi(c)]$

[9] Had we retained the parameter k, this would take the form $\sin \Pi(a)e^{f(a)/k} = \sin \Pi(b)e^{f(b)/k} = \sin \Pi(c)e^{f(c)/k}$.

Proof. Combining the last two equations yields

$$\sin \Pi(a) e^{f(a)} = \sin \Pi(b) e^{f(b)} = \sin \Pi(c) e^{f(c)}. \quad \blacksquare$$

We shall now determine the nature of our triangle invariant. If $\triangle ABC$ is an arbitrary rectilinear right triangle with side lengths a, b, and c, the invariant tells us that $\sin\Pi(a)e^{f(a)} = \sin\Pi(b)e^{f(b)}$. Moreover, for any positive number x, there is a right triangle $\triangle XBZ$ with legs x and b; applying the invariant to this triangle yields $\sin\Pi(x)e^{f(x)} = \sin\Pi(b)e^{f(b)}$. That is, the function $\sin\Pi(x)e^{f(x)}$ is a *constant* function on the positive real numbers. It therefore has nothing to do with triangles after all. Next, we shall show that it assumes the constant value 1.

Claim 11. $\sin\Pi(x)e^{f(x)} = 1$ for all positive values of x.[10]

Proof. We have argued in the previous paragraph that $\sin\Pi(x)e^{f(x)}$ is a constant function. To determine which constant value it assumes, we simply take a limit as x approaches 0 through the positive reals.

As x vanishes, $\Pi(x)$ approaches $\pi/2$ (see last lines of TP 23), so $\sin\Pi(x)$ approaches 1. Moreover, $f(x)$ vanishes with x (as is obvious from the definition of f), so $e^{f(x)}$ approaches 1. Thus, $\sin\Pi(x)e^{f(x)}$ itself approaches 1 as x vanishes. Consequently, the constant value of $\sin\Pi(x)e^{f(x)}$ must be 1, as claimed. $\blacksquare$

Corollary. $e^{-f(x)} = \sin \Pi(x)$ *for all positive values of* x.[11]

The Five Gems

Having expressed $f(x)$ in terms of $\Pi(x)$, Lobachevski can finally bring the diamond, our first new trigonometric relation, to light. A series of triangle transformations will then produce four more relations, completing the set of five gems, the matrix from which all of imaginary trigonometry will eventually be born.

Therefore,

$$\sin \Pi(c) = \sin \Pi(a) \sin \Pi(b)$$
$$\sin \Pi(\beta) = \cos \Pi(\alpha) \sin \Pi(a).$$

Moreover, by transforming the letters, these equations become

$$\sin \Pi(\alpha) = \cos \Pi(\beta) \sin \Pi(b)$$
$$\cos \Pi(b) = \cos \Pi(c) \cos \Pi(\alpha)$$
$$\cos \Pi(a) = \cos \Pi(c) \cos \Pi(\beta).$$

These five equations are easy to verify. The first gem is the long-sought diamond itself. To expose it, we remove the unsightly f-function from an equation that we obtained from the semi-polished diamond. The remaining four gems are then easy to obtain.

---

[10] Retaining the parameter, this would be $\sin\Pi(x)e^{f(x)/k} = 1$ for all positive values of x.

[11] If we retain the parameter, this takes the form $e^{-f(x)/k} = \sin\Pi(x)$ for all positive x.

Claim 12. (1st Gem) $\sin\Pi(c) = \sin\Pi(a)\sin\Pi(b)$.[12]

Proof.   We know that $\sin\Pi(b) = \sin\Pi(c)e^{f(a)}$.   (Claim 9)
Hence, $\sin\Pi(b)e^{-f(a)} = \sin\Pi(c)$.
Thus, $\sin\Pi(b)\sin\Pi(a) = \sin\Pi(c)$.   (Corollary to Claim 11)

Claim 13. (2ⁿᵈ Gem) $\sin\Pi(\beta) = \cos\Pi(\alpha)\sin\Pi(a)$.

Proof.   We know that $\sin\Pi(c) = \sin\Pi(a)\sin\Pi(b)$.   (The 1ˢᵗ Gem)
Hence, $\sin\Pi(\beta) = \sin\Pi(a)\sin\Pi(\alpha')$.   (By the L-transformation)
That is, $\sin\Pi(\beta) = \sin\Pi(a)\sin(\pi/2 - \Pi(\alpha))$.   (By definition of $\Pi(\alpha')$)
Thus, $\sin\Pi(\beta) = \sin\Pi(a)\cos\Pi(\alpha)$.

Claim 14. (3ʳᵈ Gem) $\sin\Pi(\alpha) = \cos\Pi(\beta)\sin\Pi(b)$.

Proof.   We know that $\sin\Pi(\beta) = \sin\Pi(a)\cos\Pi(\alpha)$.   (The 2ⁿᵈ Gem)
Hence, $\sin\Pi(\alpha) = \cos\Pi(\beta)\sin\Pi(b)$.   (By the O-transformation)

Claim 15. (4ᵗʰ Gem) $\cos\Pi(b) = \cos\Pi(c)\cos\Pi(\alpha)$.

Proof.   We know that $\sin\Pi(\alpha) = \cos\Pi(\beta)\sin\Pi(b)$.   (The 3ʳᵈ Gem)
Hence, $\sin\Pi(b') = \cos\Pi(c)\sin\Pi(\alpha')$.   (By the L-transformation)
That is, $\cos\Pi(b) = \cos\Pi(c)\cos\Pi(\alpha)$.   (Since $\Pi(x') = \pi/2 - \Pi(x)$)

Claim 16. (5ᵗʰ Gem) $\cos\Pi(a) = \cos\Pi(c)\cos\Pi(\beta)$.

Proof.   We know that $\cos\Pi(b) = \cos\Pi(c)\cos\Pi(\alpha)$.   (The 4ᵗʰ Gem)
Hence, $\cos\Pi(a) = \cos\Pi(c)\cos\Pi(\beta)$.   (By the O-transformation)

The Gems' Present Polish:
Sufficient for the Sphere, Not for the Plane

We may interpret the five gems as a set of statements about either:

1) the generic rectilinear right triangle,
$\triangle ABC\ [a, b, c;\ \Pi(\alpha), \Pi(\beta)]$.

      or

2) the generic spherical right triangle,
$\triangle mnk\ \{\Pi(c),\ \Pi(\beta),\ \Pi(a);\ \Pi(b),\ \Pi(\alpha')\}$.[13]

As statements about rectilinear triangles, the five
gems are still unsatisfying in their present form. The
first gem, for instance, concerns the sines of the angles

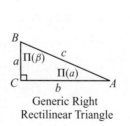

Generic Right
Rectilinear Triangle

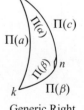

Generic Right
Spherical Triangle

---

[12] Even had we chosen to retain the parameter k, it would not occur in the five gems, so long as they remain expressed in terms of the Π-function. However, the Π-function *itself* conceals a hidden parameter. Thus, when we polish the Π's away from the five gems in TP 36, and express them Π-free notation, we shall see that the five gems do involve a parameter (whose numerical value depends upon our unit of length) after all.

[13] More precisely, $\triangle mnk$ is the generic right spherical triangle whose legs are less than $\pi/2$. (See "Right Triangle Transformation #2.)

of parallelism of the sides of $\triangle ABC$. This is certainly a vast improvement over the diamond in the rough, but we ultimately want a relation that deals directly with the sides of $\triangle ABC$ rather than with their angles of parallelism. Similar problems exist in each of the five gems when we interpret them as statements about imaginary rectilinear triangles. These problems will be resolved only after further lapidarian activities in TP 36.

In contrast, all five gems make satisfyingly direct statements about right *spherical* triangles in imaginary space. For example, when applied to $\triangle mnk$, the first gem expresses a straightforward relationship between two sides and an angle. Lobachevski takes up this theme next.

The Spherical Gems: Déjà vu

In the spherical right triangle, if the sides $\Pi(c)$, $\Pi(\beta)$, $\Pi(a)$ and opposite angles $\Pi(b)$, $\Pi(\alpha')$ are renamed a, b, c, A, B, respectively, then the preceding equations will assume forms that are known as established theorems of the ordinary spherical trigonometry of right triangles. Namely,

$$\sin(a) = \sin(c)\sin(A)$$
$$\sin(b) = \sin(c)\sin(B)$$
$$\cos(B) = \cos(b)\sin(A)$$
$$\cos(A) = \cos(a)\sin(B)$$
$$\cos(c) = \cos(a)\cos(b).$$

From these equations, we may derive those for all spherical triangles in general. Consequently, the formulae of spherical trigonometry do not depend upon whether or not the sum of the three angles in a rectilinear triangle is equal to two right angles.

If we relabel the parts of the generic spherical triangle as indicated in the figure below, and make the corresponding changes in the five gems, we obtain the following relations (*the Spherical Gems*), which hold on any imaginary spherical right triangle, $\{a, b, c; A, B\}$.

(SG1) $\sin(a) = \sin(c)\sin(A)$

(SG2) $\sin(b) = \sin(c)\sin(B)$

(SG3) $\cos(B) = \cos(b)\sin(A)$

(SG4) $\cos(A) = \cos(a)\sin(B)$

(SG5) $\cos(c) = \cos(a)\cos(b).$

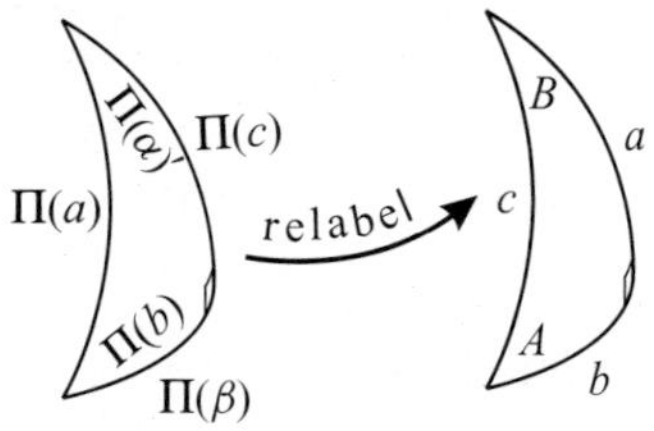

These equations would have startled Lobachevski's 19th-century audience. Having traversed for so long the increasingly alien pathways of imaginary geometry and having uncomfortably relinquished the orthodox Euclidean conception of space, their surprise must have been immense when they encountered, deep in this dark wood, five formulae that they had learned from their own schoolteachers. Unfortunately, but inevitably, the full impact of this "Young Goodman Brown"

revelation[14] is lost on most 21st century readers. We no longer study spherical trigonometry in school, and therefore we do not recognize the surprising truth before our eyes: the five equations above are basic formulae of *Euclidean* spherical trigonometry!

Thus, these five spherical trigonometric relations are *neutral* theorems; they hold in both Euclidean and imaginary space. Any further relations that we derive from them will also be neutral. Hence, when Lobachevski asserts that we can derive *every* spherical trigonometric relation from these five relations, this is equivalent to a claim that the entire subject of spherical trigonometry is neutral! That is, the formulae of spherical trigonometry are independent of the parallel postulate.

Lobachevski concludes TP 35 with this bold assertion. How do we verify it? How can we check that *every* spherical trigonometric relation derives from those given by the five spherical gems? First, we shall show that the gems hold on *all* right spherical triangles, not only those (like $\triangle mnk$) whose legs are less than $\pi/2$. Then we shall show that the gems imply that the spherical law of sines and the two spherical laws of cosines are neutral theorems. Finally, we shall argue that these three laws encompass all of spherical trigonometry: every spherical trigonometric relation is a consequence of them. Hence, if the three are neutral, all of spherical trigonometry is neutral.

The Spherical Gems Hold on All Right Spherical Triangles

The spherical gems, SG1–SG5, hold on all right spherical triangle whose legs are less than $\pi/2$ in length. A clever trick or two will quickly extend their domain to *all* right spherical triangles.

Claim 17. The spherical gems hold for all right spherical triangles whose legs are both *greater than $\pi/2$*.

Proof. Let $\triangle ABC$ be a right spherical triangle with the usual labeling,[15] in which the legs a and b are both greater than $\pi/2$. Complete the lune, as shown in the figure below, by extending the triangle's legs until they meet again at C', the point antipodal to C. Clearly, the angle at C' is also right, so $\triangle ABC'$ is a right spherical triangle with legs both less than $\pi/2$.

Applying the spherical gems to $\triangle ABC'$, we obtain the following equations:

$$\sin(\pi - a) = \sin(c)\sin(\pi - A)$$
$$\sin(\pi - b) = \sin(c)\sin(\pi - B)$$
$$\cos(\pi - B) = \cos(\pi - b)\sin(\pi - A)$$
$$\cos(\pi - A) = \cos(\pi - a)\sin(\pi - B)$$
$$\cos(c) = \cos(\pi - a)\cos(\pi - b).$$

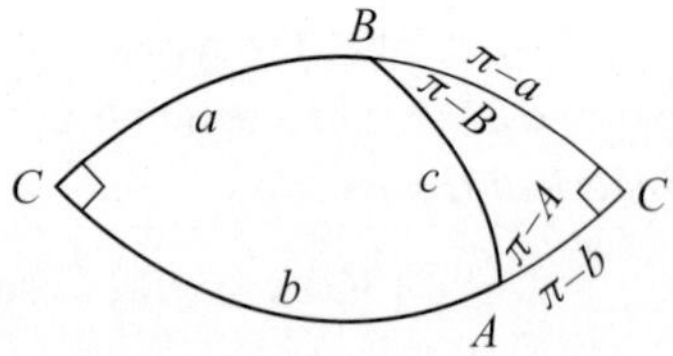

Using the identities $\sin(\pi - x) = \sin(x)$ and $\cos(\pi - x) = -\cos(x)$, these equations become the equations we wish to establish. Thus, the gems hold for $\triangle ABC$. ∎

Claim 18. The spherical gems hold for all right spherical triangles with *one leg greater than $\pi/2$, and one leg less than $\pi/2$.*

---

[14] Nathaniel Hawthorne was in fact a contemporary of Lobachevski. He published "Young Goodman Brown" in 1835.

[15] I.e., we shall denote the angles A, B, C, and the opposite sides a, b, c, respectively, with a right angle at C.

Proof. Let $\triangle ABC$ be a right spherical triangle with the usual labeling, in which $a > \pi/2$ and $b < \pi/2$. Complete the lune, as shown in the figure, extending the long leg and hypotenuse until they meet again at B', the point antipodal to B. We may now apply the gems to $\triangle AB'C$ and use an argument nearly identical to the preceding one to establish that the spherical gems hold on $\triangle ABC$, as claimed. ∎

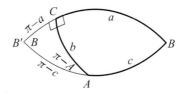

Claim 19. The spherical gems hold for all right spherical triangles which have *at least one leg equal to* $\pi/2$.

Proof. If $\triangle ABC$ is a right spherical triangle with the usual labeling and $b = \pi/2$, then A must be a "pole" for the great circle through C and B, as depicted in the figure. Thus, A is equidistant from C and B (and all other points of its "equator", the great circle through C and B), so $c = b = \pi/2$. Thus, the spherical *pons asinorum* (TP 14) yields $B = C = \pi/2$. Since an angle at a pole subtends an arc with its same measure on its equator, we have $A = a$. These equalities render each of the gems trivially true. ∎

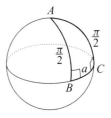

We have exhausted all cases. The five spherical gems hold for *all* right spherical triangles.

The Spherical Law of Sines is a Neutral Theorem

Claim 20. (Law of Sines) If $\triangle ABC$ is an arbitrary spherical triangle with the usual labeling, then the following relation holds, regardless of whether the parallel postulate holds:

$$\sin(A)/\sin(a) = \sin(B)/\sin(b) = \sin(C)/\sin(c).$$

Proof. If the parallel postulate holds, then so does all of Euclidean geometry and trigonometry, including the spherical law of sines. The interested reader may find a proof in older trigonometry books (written before 1950 or so), in spherical astronomy books, or in Stahl (pp. 172–3).

If the parallel postulate does not hold, then we obtain the result from SG1, as follows. Drop a perpendicular BD from B to AC (extended if necessary), and let $d = BD$. Regardless of whether D falls within AC, or to either side of this segment, we find

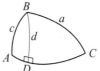

$$\sin(d) = \sin(c)\sin(A) \quad \text{(applying SG1 to the right triangle } \triangle ADB\text{)},^{16}$$
$$\sin(d) = \sin(a)\sin(C) \quad \text{(applying SG1 to the right triangle } \triangle BDC\text{)}.$$

Equating these two expressions for $\sin(d)$ yields $\sin(c)\sin(A) = \sin(a)\sin(C)$. Equivalently,

$$\sin(A)/\sin(a) = \sin(C)/\sin(c).$$

Repeating the argument, but dropping the perpendicular from C to AB, yields, by symmetry,

$$\sin(A)/\sin(a) = \sin(B)/\sin(b).$$

Combining the last two equations gives the spherical law of sines in imaginary geometry. ∎

---

[16] If D happens to fall directly upon A or C, this equation (and the one that follows) are trivially true. For example, if $D = A$, then $d = c$, and $\sin(A) = 1$, so the fact that $\sin(d) = \sin(c)\sin(A)$ is obviously true.

The First Spherical Law of Cosines is a Neutral Theorem

In our derivation of the first spherical law of cosines, we shall find the following lemma useful.

Lemma. In a right spherical triangle with the usual labeling, the following relation holds, independent of the parallel postulate:

$$\cos(B) = \tan(a)/\tan(c).$$

(Note how similar this looks to a familiar result from ordinary plane Euclidean trigonometry: $\cos(B) = a/c$.)

Proof $\cos(B) = \cos(b)\sin(A)$ (by SG 3)
$\quad\quad\quad = \cos(b)[\sin(a)/sin(c)]$ (by SG 1)
$\quad\quad\quad = [\cos(c)/\cos(a)][\sin(a)/\sin(c)]$ (by SG 5)
$\quad\quad\quad = \tan(a)/\tan(c).$ ∎

Claim 21. (First Law of Cosines) In an arbitrary spherical triangle $\triangle ABC$ with the usual labeling, the following relation holds, independent of the parallel postulate:

$$\cos(c) = \cos(a)\cos(b) + \sin(a)\sin(b)\cos(C).$$

Proof. If the parallel postulate holds, this is a classical result. If not, we prove it as follows.

If $\triangle ABC$ happens to be a right triangle, label it so that the right angle is at C. In this case, $\cos(C) = 0$, so the first law of cosines reduces to $\cos(c) = \cos(a)\cos(b)$. This is the spherical Pythagorean Theorem, which we have already proved above. (It is SG5.)

If $\triangle ABC$ is not a right triangle, we drop a perpendicular BD from B to AC, as we did in the proof of the spherical law of sines. Unfortunately, the law of cosines requires slightly different proofs (different in details, but the same in spirit) for the case in which D falls inside AC, and the case in which D falls outside AC.   ∎

Case 1 (D lies within AC). As in the figure, we let $d = BD$, $p = AD$, and $q = DC$. Then,

$\cos(c) = \cos(d)\cos(p)$ (SG5 on $\triangle ABD$)
$\quad\quad = [\cos(a)/\cos(q)]\cos(p)$ (SG5 on $\triangle BDC$)
$\quad\quad = [\cos(a)/\cos(q)]\cos(b - q)$
$\quad\quad = [\cos(a)/\cos(q)][\cos(b)\cos(q) + \sin(b)\sin(q)]$ (trig identity[17])
$\quad\quad = \cos(a)\cos(b) + \sin(b)\cos(a)\tan(q)$
$\quad\quad = \cos(a)\cos(b) + \sin(b)\cos(a)[\cos(C)\tan(a)]$ (Lemma, on $\triangle BDC$)
$\quad\quad = \cos(a)\cos(b) + \sin(a)\sin(b)\cos(C),$ as claimed.

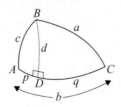

---

[17] The laws of sines and cosines are geometric theorems, so their forms will change in different geometric contexts. In contrast, trigonometric *identities* such as $\sin^2 x + \cos^2 x = 1$, or $\cos(x - y) = \cos(x)\cos(y) + \sin(x)\sin(y)$, are *not* geometric theorems. These are theorems about the trigonometric *functions* sin and cos. In other words, they are statements about numbers, rather than shapes, and consequently are independent of geometric context.

Case 2 (*D* falls outside of *AC*). Label the parts of the triangle as shown in the figure.

Then, $\cos(c) = \cos(b + r)\cos(s)$ (SG5 on $\triangle ABD$)

$\quad\quad = [\cos(b)\cos(r) - \sin(b)\sin(r)]\cos(s)$ (trig identity)

$\quad\quad = [\cos(b)\cos(r) - \sin(b)\sin(r)][\cos(a)/\cos(r)]$ (SG5 on $\triangle BCD$)

$\quad\quad = \cos(a)\cos(b) - \sin(b)\cos(a)\tan(r)$

$\quad\quad = \cos(a)\cos(b) - \sin(b)\cos(a)[\cos(\pi - C)\tan(a)]$ (Lemma, on $\triangle BCD$)

$\quad\quad = \cos(a)\cos(b) - \sin(a)\sin(b)\cos(\pi - C)$

$\quad\quad = \cos(a)\cos(b) + \sin(a)\sin(b)\cos(C),$ as claimed.

Thus, the first spherical law of cosines holds for all spherical triangles in imaginary geometry. ∎

Polar Triangles

We can establish the second law of cosines $(\cos(C) = -\cos(A)\cos(B) + \sin(A)\sin(B)\cos(c))$ with the same type of unedifying calculations that we used to prove the first cosine law. However, this approach does little to explain the striking similarity of form that the two laws exhibit. Accordingly, I shall follow a more illuminating path, which will require a brief digression on "polar triangles", a beautiful, elementary, and largely forgotten topic.

First, some terminology. Given any great circle of a sphere, we may think of it as an *equator*, which divides the sphere into two equal hemispheres. Accordingly, we shall refer to the two antipodal points on the sphere that lie furthest away from a particular equator as its *poles*. Conversely, we may think of any point on the sphere as a pole, and speak of its *equator*.

For each triangle $\triangle ABC$ on a simple sphere,[18] there is a related triangle $\triangle A'B'C'$ called its "polar triangle" (or sometimes its "dual triangle"). We shall describe its construction and a few of its most important properties.

Construction (Polar Triangle). Side BC of $\triangle ABC$ lies on an equator. Exactly one of its two poles lies in the same hemisphere as point A. (Equivalently, exactly one pole lies at a distance of less than $\pi/2$ from A.) Call this pole A'. We define the points B' and C' analogously. We call $\triangle A'B'C'$ *the polar triangle* of $\triangle ABC$.

Claim 22. Given any spherical triangle, the polar triangle of its polar triangle is identical to the original triangle.

Proof. Let $\triangle A'B'C'$ be the polar triangle of $\triangle ABC$. By definition of the polar triangle, B' is a pole of equator AC, and C' is a pole of equator AB.

---

[18] Recall that on a simple sphere, the length of any arc of a great circle equals the measure of the angle it subtends at the sphere's center.

The distance, as measured along a great circle arc, from a pole to any point on its equator is clearly $\pi/2$ (on a simple sphere), so $B'A = C'A = \pi/2$.

Since B' and C' therefore lie at a distance of $\pi/2$ from A, it follows that points B' and C' lie upon the equator of A. Equivalently, A is a pole of $B'C'$. Since the distance between A and A' is already known to be less than $\pi/2$, point A is a vertex of the polar triangle of $\triangle A'B'C'$.

Similarly, B and C are vertices of the polar triangle of $\triangle A'B'C'$.

Thus, the polar triangle of the polar triangle of $\triangle ABC$ is $\triangle ABC$, as claimed. ∎

Hence, polar triangles come in pairs. In each such pair, there is a remarkable relationship between their parts. Namely, each part (side or angle) of either triangle is the *supplement* of a related part of the other triangle. In particular, if $\triangle ABC$ and $\triangle A'B'C'$ are a polar pair, each labeled in the usual way (side a opposite angle A, etc.), then

$$a + A' = \pi \quad b + B' = \pi \quad c + C' = \pi \quad A + a' = \pi \quad B + b' = \pi \quad C + c' = \pi.$$

Before proving that these relationships hold, we need to establish a simple lemma.

Lemma. *On a simple sphere, any angle has the same measure as the arc that it subtends on its vertex's equator.*

Proof. A picture is worth a thousand words. In the picture at right, it is clear that angle θ and length l are directly proportional. Since the "full angle" at the pole is 2π, and the "full length" (i.e., the equator's circumference) is 2π (by definition of a simple sphere), the constant of proportionality must be 1. That is, $\theta = l$, as claimed. ∎

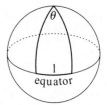

Claim 23. Given a spherical triangle $\triangle ABC$ and its polar triangle $\triangle A'B'C'$, each part (side/angle) of either triangle is the supplement of the (angle/side) lying opposite its corresponding part on the other triangle.

(That is, the following are supplementary pairs: $\{a, A'\}, \{b, B'\}, \{c, C'\}, \{a', A\}, \{b', B\}, \{c', C\}$.)

Proof. First, we show that B and b' are supplementary:

Extend the sides of $\angle B$ (if necessary) until they cut the great circle containing side b'. Let D and E be the points of intersection, as in the figure. By construction, D lies on the equator containing BC, whose pole is A'.

Hence, arc $DA' = \pi/2$.

Similarly, arc $EC' = \pi/2$.

Substituting these values into the equation $DA' + EC' = A'C' + ED$ (see the figure) yields $\pi = b' + ED$.

Thus, since $ED = B$ (by the Lemma), it follows that $\pi = b' + B$. That is, B and b' are supplementary, as claimed.

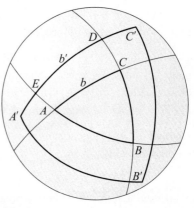

Similarly, A and a' must be supplementary, as must be C and c'.

We have demonstrated that the angles of a spherical triangle are supplementary to the sides opposite their corresponding angles on the polar triangle. Applying this result to $\triangle A'B'C'$ (whose polar triangle is $\triangle ABC$, by Proposition 1) tells us that A' and a are supplementary, as are B' and b, as well as C' and c.

This completes the proof.[19] ∎

The Second Spherical Law of Cosines is a Neutral Theorem

The preceding result on polar triangles is a neutral theorem, and thus we may use it in imaginary geometry. It affords a spectacularly simple proof of the second spherical law of cosines.

Claim 24. (Second Law of Cosines) In an arbitrary spherical triangle $\triangle ABC$ with the usual labeling, the following relation holds, independent of the parallel postulate:

$$\cos(C) = -\cos(A)\cos(B) + \sin(A)\sin(B)\cos(c).$$

Proof. Let $\triangle ABC$ be a spherical triangle in either Euclidean or imaginary geometry. Let $\triangle A'B'C'$ be its polar triangle. Applying the first law of cosines (Claim 21) to the polar triangle, we find that

$$\cos(c') = \cos(a')\cos(b') + \sin(a')\sin(b')\cos(C').$$

Using Claim 23 to express the parts of $\triangle A'B'C'$ in terms of the parts of $\triangle ABC$, this last equation becomes

$$\cos(\pi - C) = \cos(\pi - A)\cos(\pi - B) + \sin(\pi - A)\sin(\pi - B)\cos(\pi - c).$$

That is,

$$-\cos(C) = \cos(A)\cos(B) - \sin(A)\sin(B)\cos(c).$$

Multiplying both sides by -1 yields

$$\cos(C) = -\cos(A)\cos(B) + \sin(A)\sin(B)\cos(c). \qquad \blacksquare$$

Spherical Trigonometry is a Neutral Subject

We have now demonstrated that the laws of sines and cosines are neutral theorems. It remains only to argue that all spherical trigonometric relations are consequences of these theorems, and hence all of spherical trigonometry is independent of the parallel postulate. This amounts to little more than a review of high school geometry and trigonometry.

One of the basic problems of geometry is to recognize when two triangles are congruent.

In the Euclidean plane, this problem is solved by the congruence criteria (SSS, SAS, ASA, AAS), each of which specifies a trio of data sufficient to fix a triangle's size and shape. Any such trio determines *all six* of the triangle's parts. For example, given three sides of a triangle, SSS guarantees us that its three angles are, in theory, completely determined. To find their numerical values in practice (i.e. to "solve the triangle"), one must go beyond *The Elements* and turn to trigonometry. In its literal sense of triangle measurement, trigonometry is simply an adjunct to

---

[19] In this proof, I have assumed that D and E fall between A' and C' (i.e. that they actually lie on side $A'C'$ of the polar triangle). This need not happen. If it does not, then obvious adjustments in the proof can be made to salvage it. In the spirit of Euclid, I will refrain from explicitly proving these trivial variations.

the congruence criteria. Its tools are "trigonometric relations", equations that relate the various parts of triangles to one another. In fact, one can get by with only two relations in the Euclidean plane.

The laws of cosines and sines suffice to solve every problem of Euclidean plane trigonometry.

As a practical demonstration of this claim, note that if we know the three sides of the triangle (SSS), we can use the law of cosines three times in succession to determine its angles. Given two sides and their included angle (SAS), the law of cosines will yield the third side, reducing the problem to the already solved SSS case. When two angles are known (ASA or AAS), we can use Euclid I.32 (the angle sum of any triangle is π) to find the third angle and then the law of sines to produce a second side, reducing the problem to SAS. Thus, the laws of cosines and sines do indeed completely encapsulate plane trigonometry. Similarly,

The two spherical laws of cosines and the spherical law of sines suffice to solve every problem of spherical trigonometry.

As in plane trigonometry, the (first) law of cosines suffices to solve a triangle in SSS or SAS case. When two angles are known (ASA or AAS), we must use the *second* law of cosines to find the third angle (Euclid I.32 does not hold on the sphere), and then apply the law of sines to find a second side, reducing the problem to SAS. Finally, we can handle AAA, which is a congruence criterion on the sphere, by using the second law of cosines three times in succession to determine the sides. Hence, the spherical law of sines and two spherical laws of cosines encapsulate Euclidean spherical trigonometry, as claimed.

Consequently, the neutrality of the spherical laws of sines and cosines guarantees the neutrality of spherical trigonometry.

An Aside: Why the Spherical Gems Would Have Looked Familiar to 19th-Century Readers

As I mentioned earlier, readers raised with spherical trigonometry would recognize the spherical interpretations of the five gems as old schoolfriends. SG5 is the spherical Pythagorean theorem. But what of the first four? To explain why these would have been familiar requires another return to high school mathematics.

Although in theory the laws of sines and cosines suffice to solve all trigonometric problems in the Euclidean plane, we often forgo them in practice in favor of the humbler formulae specific to *right*-triangle trigonometry. These are useful in practice because we may decompose any non-right triangle into two right triangles by dropping an altitude from an appropriate vertex.

Right-triangle trigonometric relations are merely special cases of the laws of cosines and sines. In the Euclidean plane, they are simple: in a right triangle, the law of cosines reduces to the Pythagorean Theorem, while the law of sines reduces to a compact set of rules often summarized by the equations

$$\text{Sin} = \text{Opposite/Hypotenuse}, \quad \text{Cos} = \text{Adjacent/Hypotenuse}, \quad \text{Tan} = \text{Opposite/Adjacent}.$$

Many students learn these today with the mnemonic name **SOH CAH TOA**, the ancient god of plane trigonometry.

However, on the sphere, right triangle trigonometry is not quite so simple. There, SOH CAH TOA's monotheistic rule is replaced by a pantheon of ten equations, each governing a specific combination of parts of the right spherical triangle.[20] The five gems, in their spherical interpretation, are five of these ten equations that govern right-triangle spherical trigonometry.

A Quibble

It is tempting to conclude that if we pluck a simple sphere out of Euclidean space and thrust it into imaginary space, its trigonometric formulae remain unchanged. This is almost, but not quite correct. Because we have been working with a *simple* sphere throughout this proposition, we should state what Lobachevski has proved more precisely: trigonometry on a *simple* sphere agrees in both spaces.

A sphere is simple if and only if its great circles have circumference 2π. Thus, in Euclidean space, where $C = 2\pi r$, a sphere is simple if and only if its radius is 1. However, in imaginary space, circumference is given by $C = 2\pi \sinh(r)$,[21] so a sphere is simple if and only if its radius is $\sinh^{-1}(1)$. Thus, a simple Euclidean sphere thrust into imaginary space (or vice versa) would cease to be simple in its new context. However, the changes in the formulae are insignificant. Trigonometry on an imaginary sphere of radius r will always agree with trigonometry of some Euclidean sphere; namely, the sphere of radius $\sinh(r)$. This agreement with some Euclidean sphere is the important point. The differences between one Euclidean sphere and another are trivial.

---

[20] Each equation asserts the precise relationship between a particular trio of a right triangle's five variable parts (its sides and non-right angles). There are ten equations precisely because there are ten ways to choose trios from amongst a set of five. There is, incidentally, a mnemonic device for recalling the ten equations, known as Napier's Rule.

[21] Lobachevski proves this elsewhere (in *Pangeometry*, for example), but not in *The Theory of Parallels*. Note that the formula for circumference implies that π loses its usual geometric meaning in imaginary geometry: the ratio of circumference to diameter is *not* a constant in this setting.

Theory of Parallels 36

Having revealed the structure of spherical trigonometry, Lobachevski returns to the plane to settle some old business: twenty propositions after introducing it, he finally derives an explicit formula for the Π-function.

Lobachevski begins by deriving a pair of relations among $\triangle ABC$'s parts.

First Relation: $\Pi(b) = \Pi(\alpha) + \Pi(c + \beta)$.

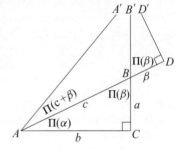

We now return to the rectilinear right triangle $\triangle ABC$ with sides a, b, c, and opposite angles $\Pi(\alpha)$, $\Pi(\beta)$, $\pi/2$. Extend this triangle's hypotenuse beyond B to a point D at which $BD = \beta$, and erect a perpendicular DD' from BD. By construction, DD' is parallel to BB', the extension of side a beyond B. Finally, draw AA' parallel to DD'; it will be parallel to CB' as well (TP 25).

From this, we have $\angle A'AC = \Pi(b)$ and $\angle A'AD = \Pi(c + \beta)$, from which it follows that

$$\Pi(b) = \Pi(\alpha) + \Pi(c + \beta).$$

Lobachevski draws ray CB and two rays parallel to it: one emanating from A, the other perpendicular to the hypotenuse. Each is uniquely determined.

The derivation of the relation in this passage is self-explanatory, but it does depend upon the fact that $\Pi(b) > \Pi(\alpha)$. Lobachevski does not bother to prove this, since it is so easy to justify. Among the pencil of rays that emanate from A, each ray either cuts or does not cut CB. Since AA' does not cut it, while AB does, we know that $\angle A'AC > \angle BAC$. That is, $\Pi(b) > \Pi(\alpha)$, as claimed.

Second Relation: $\Pi(c - \beta) = \Pi(\alpha) + \Pi(b)$.

The construction here is almost identical to the one above, except that we draw the rays parallel to BC instead of CB. The derivation of the second equation is slightly messier, however, as it requires three separate cases, according to whether β is less than, equal to, or greater than c. He begins with $\beta < c$.[1]

Now let E be the point on ray BA for which $BE = \beta$. Erect the perpendicular EE' to AB, and draw AA'' parallel to it. Line BC', the extension of side a beyond C, will be a third parallel.

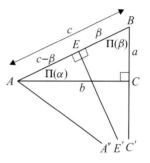

If $\beta < c$, as in the figure, we see that $\angle CAA'' = \Pi(b)$ and

$\angle EAA'' = \Pi(c - \beta)$, from which it follows that

$$\Pi(c - \beta) = \Pi(\alpha) + \Pi(b).$$

In fact, this last equation remains valid even when $\beta = c$, or $\beta > c$.

The two remaining cases ($\beta = c$ and $\beta > c$) are straightforward, although Lobachevski makes what may appear to be inappropriate references to TP 23 in the midst of establishing them. In fact, these are not references to the *theorem* in TP 23 ("For any given angle α, there is a line p such that $\Pi(p) = \alpha$."), but to two facts about the Π-function that Lobachevski first mentions in TP 23: $\Pi(0) = \pi/2$, and $\Pi(-x) = \pi - \Pi(x)$, by definition, for any x.

If $\beta = c$ (see the figure at the left below), the perpendicular AA' erected upon AB is parallel to BC, and hence to CC', from which it follows that $\Pi(\alpha) + \Pi(b) = \pi/2$. Moreover, $\Pi(c - \beta) = \pi/2$ (TP 23).

If $\beta > c$ (see the figure at the right below), E falls beyond point A. In this case, we have $\angle EAA'' = \Pi(c - \beta)$, from which it follows that

$$\Pi(\alpha) + \Pi(b) = \pi - \Pi(\beta - c) = \Pi(c - \beta) \text{ (TP 23)}.$$

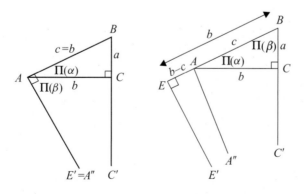

---

[1] I have taken the liberty of changing some of Lobachevski's notation in the following passage. My E is actually a *second* point D in the manuscript, while my AA'' is a second AA'. Using the same symbol to denote distinct objects that play similar roles was common practice in 19th-century mathematics.

Combining the Relations

With a little algebraic manipulation, Lobachevski will now combine the two relations into a new form, which is related to the 4th gem from TP 35 ($\cos \Pi(b) = \cos \Pi(c) \cos \Pi(\alpha)$).

Combining the two equations yields

$$2\Pi(b) = \Pi(c - \beta) + \Pi(c + \beta)$$
$$2\Pi(\alpha) = \Pi(c - \beta) - \Pi(c + \beta),$$

from which follows

$$\frac{\cos \Pi(b)}{\cos \Pi(\alpha)} = \frac{\cos[\frac{1}{2}\Pi(c - \beta) + \frac{1}{2}\Pi(c + \beta)]}{\cos[\frac{1}{2}\Pi(c - \beta) - \frac{1}{2}\Pi(c + \beta)]}.$$

Claim 1. In the generic right rectilinear triangle $\triangle ABC$, the following holds:

$$\frac{\cos \Pi(b)}{\cos \Pi(\alpha)} = \frac{\cos[\frac{1}{2}\Pi(c - \beta) + \frac{1}{2}\Pi(c + \beta)]}{\cos[\frac{1}{2}\Pi(c - \beta) - \frac{1}{2}\Pi(c + \beta)]} \quad . \tag{1}$$

Proof. Subtracting the second relation from the first and solving for $\Pi(b)$ yields a "doubling formula" for $\Pi(b)$:

$$2\Pi(b) = \Pi(c - \beta) + \Pi(c + \beta).$$

Adding the corresponding sides of the two relations yields a "doubling formula" for $\Pi(\alpha)$:

$$2\Pi(\alpha) = \Pi(c - \beta) - \Pi(c + \beta).$$

Hence,

$$\cos \Pi(b) = \cos(1/2 \cdot 2\Pi(b))$$
$$= \cos(1/2\Pi(c - \beta) + 1/2\Pi(c + \beta)) \quad \text{(by the first doubling formula)},$$

and

$$\cos \Pi(\alpha) = \cos(1/2 \cdot 2\Pi(\alpha))$$
$$= \cos(1/2\Pi c - \beta) - 1/2\Pi(c + \beta)) \quad \text{(by the second doubling formula)}.$$

Therefore,

$$\frac{\cos \Pi(b)}{\cos \Pi(\alpha)} = \frac{\cos[\frac{1}{2}\Pi(c - \beta) + \frac{1}{2}\Pi(c + \beta)]}{\cos[\frac{1}{2}\Pi(c - \beta) - \frac{1}{2}\Pi(c + \beta)]}, \quad \text{as claimed.} \quad \blacksquare$$

A Tangy Equation

Thus far in TP 36, Lobachevski has yet to use any substantial results from imaginary geometry. This will now change.

Using the substitution

$$\frac{\cos \Pi(b)}{\cos \Pi(\alpha)} = \cos \Pi(c) \text{ (from TP 35)}$$

yields

$$\tan^2 \left(\frac{\Pi(c)}{2} \right) = \tan \left(\frac{\Pi(c - \beta)}{2} \right) \tan \left(\frac{\Pi(c + \beta)}{2} \right).$$

The "substitution" is the fourth gem ($\cos \Pi(b) = \cos \Pi(c) \cos \Pi(\alpha)$). We shall derive the other equation in this passage with the help of some trigonometric gymnastics, and the following oft-forgotten identity.

$$\tan^2 \left(\frac{x}{2} \right) = \frac{1 - \cos x}{1 + \cos x}. \quad \text{(Half-angle identity for tangent)}$$

Claim 2. If c is the hypotenuse of a rectilinear right triangle in imaginary geometry, and $\Pi(\beta)$ the measure of one of its acute angles, then the following relation holds.

$$\tan^2 \left(\frac{\Pi(c)}{2} \right) = \tan \left(\frac{\Pi(c - \beta)}{2} \right) \tan \left(\frac{\Pi(c + \beta)}{2} \right). \tag{2}$$

Proof.

$$\tan^2 \left(\frac{\Pi(c)}{2} \right) = \frac{1 - \cos \Pi(c)}{1 + \cos \Pi(c)} \quad \text{(half-angle identity for tangent)}$$

$$= \frac{1 - \dfrac{\cos \Pi(b)}{\cos \Pi(\alpha)}}{1 + \dfrac{\cos \Pi(b)}{\cos \Pi(\alpha)}} \quad \text{(4}^{\text{th}}\text{ gem: TP 35 Notes, Claim 15)}$$

$$= \frac{1 - \dfrac{\cos \left[\frac{1}{2}\Pi(c - \beta) + \frac{1}{2}\Pi(c + \beta) \right]}{\cos \left[\frac{1}{2}\Pi(c - \beta) - \frac{1}{2}\Pi(c + \beta) \right]}}{1 + \dfrac{\cos \left[\frac{1}{2}\Pi(c - \beta) + \frac{1}{2}\Pi(c + \beta) \right]}{\cos \left[\frac{1}{2}\Pi(c - \beta) - \frac{1}{2}\Pi(c + \beta) \right]}} \quad \text{(equation (1))}$$

$$= \frac{\cos \left[\frac{1}{2}\Pi(c - \beta) - \frac{1}{2}\Pi(c + \beta) \right] - \cos \left[\frac{1}{2}\Pi(c - \beta) + \frac{1}{2}\Pi(c + \beta) \right]}{\cos \left[\frac{1}{2}\Pi(c - \beta) - \frac{1}{2}\Pi(c + \beta) \right] + \cos \left[\frac{1}{2}\Pi(c - \beta) + \frac{1}{2}\Pi(c + \beta) \right]}. \text{(algebra)}$$

If we expand all four cosines in this expression with the addition/subtraction formulas for cosine ($\cos(A \pm B) = \cos A \cos B \mp \sin A \sin B$) and simplify the result, it becomes

$$\frac{2 \sin \left[\frac{1}{2}\Pi(c - \beta) \right] \sin \left[\frac{1}{2}\Pi(c + \beta) \right]}{2 \cos \left[\frac{1}{2}\Pi(c - \beta) \right] \cos \left[\frac{1}{2}\Pi(c + \beta) \right]}$$

$$= \tan \left[\frac{1}{2}\Pi(c - \beta) \right] \tan \left[\frac{1}{2}\Pi(c + \beta) \right].$$

That is,

$$\tan^2 \left(\frac{\Pi(c)}{2} \right) = \tan \left(\frac{\Pi(c - \beta)}{2} \right) \tan \left(\frac{\Pi(c + \beta)}{2} \right), \text{ as claimed. } \blacksquare$$

The Fundamental Formula

> "This is certainly one of the most remarkable formulas in all of mathematics, and it is astonishing how few mathematicians know it."
>
> —Marvin Greenberg[2]

It is now but a short step to the equation that will finally reveal the precise nature of the Π-function:

$$\tan\left(\frac{\Pi(x)}{2}\right) = e^{-x}. \tag{3}$$

We are justified in referring to this as the *fundamental formula* of imaginary geometry since it yields an explicit expression for the all-pervasive angle of parallelism: $(x) = 2\arctan(e^{-x})$.

Because the angle $\Pi(\beta)$ at B may have any value between 0 and $\pi/2$, β itself can be any number between 0 and ∞. By considering the cases in which $\beta = c, 2c, 3c$, *etc.*, we may deduce that for all positive values of r,[3]

$$\tan^r\left(\frac{\Pi(c)}{2}\right) = \tan\left(\frac{\Pi(rc)}{2}\right).$$

If we view r as the ratio of two values x and c, and assume that $\cot(\Pi(c)/2) = e^c$, we find that for all values of x, whether positive or negative,

$$\tan\left(\frac{\Pi(x)}{2}\right) = e^{-x},$$

where e is an indeterminate constant, which is larger than 1, since $\Pi(x) = 0$ when $x = \infty$.

Since the unit with which we measure lengths may be chosen at will, we may choose it so that e is the base of the natural logarithm.

Lobachevski's proof of the fundamental formula is sketchy and somewhat unclear. I shall present an alternate proof, which is complete and transparent, based on the functional equation that we have already used to establish an exponential relationship in TP 33 (between corresponding arcs on concentric horocycles). At the end of the notes, however, I will return to Lobachevski's method, and show how to expand his sketch into a full proof of the fundamental formula. The reader may then decide for himself which method of proof he prefers.

Theorem. $\tan\left(\frac{\Pi(x)}{2}\right) = e^{-x}$ *for any real value of* x.[4]

---

[2] Greenberg, p. 323.

[3] Where I have r, Lobachevski uses the symbol n. Whatever one calls it, it stands for any positive *real* number. I have switched to r so as to conform with the convention of reserving n for natural numbers.

[4] If we retain the parameter k, the right-hand side of this expression becomes $e^{-x/k}$, which is shown explicitly in this proof.

Proof. If a right rectilinear triangle has hypotenuse c and an acute angle $\Pi(\beta)$, then we know that:

$$\tan^2\left(\frac{\Pi(c)}{2}\right) = \tan\left(\frac{\Pi(c-\beta)}{2}\right)\tan\left(\frac{\Pi(c+\beta)}{2}\right). \qquad (2)$$

Note, however, that for *any* positive c and β, there is a right triangle with hypotenuse c and an angle $\Pi(\beta)$.[5] Thus, (2) holds for any positive values of c and β. In particular, if x and y are arbitrary positive reals, we may write

$$\tan^2\left(\frac{\Pi(x)}{2}\right) = \tan\left(\frac{\Pi(x-y)}{2}\right)\tan\left(\frac{\Pi(x+y)}{2}\right), \quad \text{(letting } c = x \text{ and } \beta = y \text{ in (2))}$$

and

$$\tan^2\left(\frac{\Pi(y)}{2}\right) = \tan\left(\frac{\Pi(y-x)}{2}\right)\tan\left(\frac{\Pi(y+x)}{2}\right) \quad \text{(letting } c = y \text{ and } \beta = x \text{ in (2)).}$$

Multiplying these, we obtain

$$\tan^2\left(\frac{\Pi(x)}{2}\right)\tan^2\left(\frac{\Pi(y)}{2}\right)$$
$$= \tan\left(\frac{\Pi(x-y)}{2}\right)\tan\left(\frac{\Pi(x+y)}{2}\right)\tan\left(\frac{\Pi(y-x)}{2}\right)\tan\left(\frac{\Pi(x+y)}{2}\right)$$
$$= \tan^2\left(\frac{\Pi(x+y)}{2}\right)\tan\left(\frac{\Pi(x-y)}{2}\right)\tan\left(\frac{\Pi(y-x)}{2}\right)$$
$$= \tan^2\left(\frac{\Pi(x+y)}{2}\right)\tan\left(\frac{\Pi(x-y)}{2}\right)\tan\left(\frac{\pi-\Pi(x-y)}{2}\right) \quad \text{(definition of } \Pi(-x))$$
$$= \tan^2\left(\frac{\Pi(x+y)}{2}\right)\tan\left(\frac{\Pi(x-y)}{2}\right)\cot\left(\frac{\Pi(x-y)}{2}\right)$$
$$= \tan^2\left(\frac{\Pi(x+y)}{2}\right).$$

Taking square roots yields

$$\tan\left(\frac{\Pi(x+y)}{2}\right) = \tan\left(\frac{\Pi(x)}{2}\right)\tan\left(\frac{\Pi(y)}{2}\right).$$

Thus, $f(x) = \tan^2\left(\frac{\Pi(x)}{2}\right)$ satisfies the functional equation $f(x+y) = f(x)f(y)$.

Hence, by Claim 3 in the notes to TP 33, we must have

$$\tan^2\left(\frac{\Pi(x)}{2}\right) = a^x,$$

for all $x \geq 0$, where $a = f(1)$. In fact, the identity holds for all negative arguments as well: since $\Pi(-x) = \pi - \Pi(x)$, we have

$$\tan^2\left(\frac{\Pi(-x)}{2}\right) = \tan^2\left(\frac{\pi-\Pi(x)}{2}\right) = \frac{1}{\tan^2\left(\frac{\Pi(x)}{2}\right)} = \frac{1}{a^x} = a^{-x}.$$

---

[5] Let AB be a segment of length c. Draw ray BB' such that $\angle B'BA = \Pi(\beta)$. Drop a perpendicular AC from A to BB'. $\triangle ABC$ is the required triangle.

Thus, $\tan^2\left(\frac{\Pi(x)}{2}\right) = a^x$ holds for *all* real values of x. Note that the numerical value of

$$a = \tan^2\left(\frac{\Pi(1)}{2}\right)$$

depends upon the unit of length, since, for example, $\Pi(1\text{ millimeter})$ is not the same as $\Pi(1\text{ light-year})$. In fact, if we choose the unit of length judiciously, we may endow $\Pi(1)$ with any value between 0 and $\pi/2$, and thus we may endow a with any value between 0 and 1. For each value a in this range, there is a unique $k > 0$ such that $e^{-1/k} = a$. Hence, we may write

$$\tan^2\left(\frac{\Pi(x)}{2}\right) = e^{-x/k},$$

where $k > 0$ is a parameter, whose numerical value depends upon the unit of length. Lobachevski selects his unit of length to be that which makes $k = \frac{1}{2}$, and thus writes

$$\tan\left(\frac{\Pi(x)}{2}\right) = e^{-x},$$

as claimed.   ■

The Rosetta Stone

The "five gems" of TP 35 are remarkable inasmuch as they simultaneously describe trigonometric relations on a spherical triangle and on a rectilinear triangle. We found that these five equations,

$$\sin\Pi(c) = \sin\Pi(a)\sin\Pi(b)$$
$$\sin\Pi(\beta) = \cos\Pi(\alpha)\sin\Pi(a)$$
$$\sin\Pi(\alpha) = \cos\Pi(\beta)\sin\Pi(b)$$
$$\cos\Pi(b) = \cos\Pi(c)\cos\Pi(\alpha)$$
$$\cos\Pi(a) = \cos\Pi(c)\cos\Pi(\beta)$$

were immediately comprehensible as statements about the sides and angles of our generic spherical right triangle $\{\Pi(c), \Pi(\beta), \Pi(a); \Pi(b), \Pi(\alpha')\}$, and they led to a complete understanding of spherical trigonometry in imaginary space. In contrast, thus far we have been unable to fully grasp the rectilinear interpretation of these equations, as they include the terms $\Pi(a)$, $\Pi(b)$, and $\Pi(c)$, none of which have simple interpretations as sides or angles on our generic rectilinear right triangle $[a, b, c; \Pi(\alpha), \Pi(\beta)]$.

However, now that we have an explicit formula for the Π-function, we can polish it away from the five gems, yielding simple statements about the sides a, b, and c, instead of their angles of parallelism, in which we have little interest when trying to solve trigonometric problems.

Rather than applying the fundamental formula directly to the five gems, we shall produce some simple substitutions for $\sin\Pi(x)$, $\cos\Pi(x)$, and $\tan\Pi(x)$, and use these instead.

Claim 3. $\tan \Pi(x) = \frac{1}{\sinh x}$ for all x.[6]

Proof.

$$\tan \Pi(x) = \frac{\sin \left(2 \frac{\Pi(x)}{2} \right)}{\cos \left(2 \frac{\Pi(x)}{2} \right)} \quad \text{(defn. of tangent)}$$

$$= \frac{2 \sin \left(\frac{\Pi(x)}{2} \right) \cos \left(\frac{\Pi(x)}{2} \right)}{\cos^2 \left(\frac{\Pi(x)}{2} \right) - \sin^2 \left(\frac{\Pi(x)}{2} \right)} \quad \text{(double angle formulae)}$$

$$= \frac{2 \tan \left(\frac{\Pi(x)}{2} \right)}{1 - \tan^2 \left(\frac{\Pi(x)}{2} \right)} \quad \text{(dividing top and bottom by } \cos^2(\Pi(x)/2))$$

$$= \frac{2e^{-x}}{1 - e^{-2x}} \quad \text{(the fundamental formula)}$$

$$= \frac{2}{e^x - e^{-x}} \quad \text{(multiplying top and bottom by } e^x)$$

$$= \frac{1}{\sinh x}. \quad \text{(defn. of sinh)} \quad \blacksquare$$

Claim 4. $\sin \Pi(x) = \frac{1}{\cosh x}$ for all x.[7]

Proof.

$$\cosh^2(x) = 1 + \sinh^2(x) \quad \text{(hyperbolic trig. identity)}$$
$$= 1 + \cot^2 \Pi(x) \quad \text{(Claim 3)}$$
$$= \csc^2 \Pi(x) \quad \text{(trig. identity)}$$
$$= 1/\sin^2 \Pi(x). \quad \text{(defn. of csc).}$$

Solving for $\sin \Pi(x)$ yields the desired result. $\blacksquare$

Claim 5. $\cos \Pi(x) = \tanh x$ for all x.[8]

Proof.

$$\cos \Pi(x) = \sin \Pi(x)/\tan \Pi(x)$$
$$= (1/\cosh(x))/(1/\sinh(x)) \quad \text{(Claims 3 and 4)}$$
$$= \sinh(x)/\cosh(x)$$
$$= \tanh(x). \quad \blacksquare$$

---

[6] If we retain the parameter k, this becomes $\tan \Pi(x) = 1/\sinh(x/k)$, as is easily seen by making the appropriate changes in the proof.

[7] With the parameter, $\sin \Pi(x) = 1/\cosh(x/k)$.

[8] With the parameter, $\cos \Pi(x) = \tanh(x/k)$.

The Five Gems Revisited: the Planar Interpretation

With these three substitutions, we may read the five gems as simple statements about our generic rectilinear right triangle $\triangle ABC$ [a, b, c; $\Pi(\alpha)$, $\Pi(\beta)$]. For example, the first gem is $\sin \Pi(c) = \sin \Pi(a) \sin \Pi(b)$. Applying the substitution in Claim 4 yields

$$\cosh(c) = \cosh(a) \cosh(b),$$

which is the *Pythagorean theorem in the imaginary plane*, since it relates the three sides of any rectilinear right triangle in imaginary geometry.

The second gem is $\sin \Pi(\beta) = \cos \Pi(\alpha) \sin \Pi(a)$. We need not eliminate $\Pi(\alpha)$ or $\Pi(\beta)$ from it, since these are angles in $\triangle ABC$. In contrast, we want a statement that involves a, not $\Pi(a)$. Consequently, we can eliminate $\sin\Pi(a)$ to obtain $\cos \Pi(\alpha) = \cosh(a) \sin \Pi(\beta)$. This will look neater if we use the customary symbols A and B for the angles, instead of $\Pi(\alpha)$ and $\Pi(\beta)$. After this cosmetic change, the second gem becomes

$$\cos A = \cosh(a) \sin(B).$$

After applying the same strategies to the remaining equations, the five gems assume their rectilinear interpretations (i.e., *the **Rectilinear Gems***)[9]:

(RG)	$\cosh(c) = \cosh(a) \cosh(b)$	
(RG2)	$\cos(A) = \cosh(a) \sin(B)$	
(RG3)	$\cos(B) = \cosh(b) \sin(A)$	
(RG4)	$\tanh(b) = \tanh(c) \cos(A)$	
(RG5)	$\tanh(a) = \tanh(c) \cos(B)$.	

Just as the relationships among the parts of a right spherical triangle are described by ten rules (Napier's rules), the parts of a right rectilinear triangle in the imaginary plane are related by ten analogous rules. We have just found five of them. The laws of cosines and sines for imaginary plane trigonometry are, in fact, implicit in these five equations; hence, all of imaginary plane trigonometry is implicit in them. In TP 37, Lobachevski will make this explicit, deriving the fundamental trigonometric relationships of the imaginary plane, thus fulfilling his promise at the end of TP 22.

Curiously, Lobachevski does not bother to translate the $\Pi(a)$, $\Pi(b)$, and $\Pi(c)$ terms into the language of hyperbolic functions of a, b, and c. Since the trigonometric relations are so much easier to comprehend after this translation, I shall perform this routine task for him in the notes to TP 37. For now, I conclude this section with a summary of the translation process for future reference.

Translation Summary

To translate Lobachevski's Π-laden equations about a generic rectilinear triangle (with sides a, b, c, and opposite angles $\Pi(\alpha)$, $\Pi(\beta)$, and $\Pi(\gamma)$) into Π-free equations about the generic rectilinear triangle (with sides a, b, c, and opposite angles A, B, and C), we use the following substitutions:

---

[9] If the parameter k is retained, the side lengths appearing in the rectilinear gems (a, b, c) would each be divided by k. Thus, for example, RG2 would become $\cos(A) = \cosh(a/k) \sin(B)$.

$$\sin \Pi(x) = \frac{1}{\cosh x}, \quad \cos \Pi(x) = \tanh x, \quad \tan \Pi(x) = \frac{1}{\sinh x},^{10}$$

$$\Pi(\alpha) = A, \quad \Pi(\beta) = B, \quad \Pi(\gamma) = C.$$

Appendix: Lobachevski's Derivation of the Fundamental Formula

Lobachevski's derivation requires more work than the functional equation approach did. The key is to prove that the following identity holds for all positive real numbers c and r:

$$\tan^r \left(\frac{\Pi(c)}{2} \right) = \tan \left(\frac{\Pi(rc)}{2} \right). \tag{4}$$

Lobachevski's hint (to let $\beta = c, 2c, 3c$, etc. in equation (2)) suffices only to establish the identity when r is a natural number. We shall begin here, and then extend the result to all positive values of r.

Claim 6. The identity $\tan^r \left(\frac{\Pi(c)}{2} \right) = \tan \left(\frac{\Pi(rc)}{2} \right)$ holds whenever r is a natural number (and c is any positive real).

Proof. The identity obviously holds when $r = 0$ or 1. Setting $\beta = c$ in (2) shows that it holds when $r = 2$. Setting $\beta = 2c$ in (2) yields

$$\tan^2 \left(\frac{\Pi(c)}{2} \right) = \tan \left(\frac{\Pi(-c)}{2} \right) \tan \left(\frac{\Pi(3c)}{2} \right)$$

$$= \tan \left(\frac{\pi - \Pi(c)}{2} \right) \tan \left(\frac{\Pi(3c)}{2} \right)$$

$$= \cot \left(\frac{\Pi(c)}{2} \right) \tan \left(\frac{\Pi(3c)}{2} \right).$$

Multiplying both sides of this equation by $\tan(\Pi(c)/2)$ establishes the identity when $r = 3$.

We can base a routine induction argument on this last case to show that identity (4) holds when r is any natural number. I shall omit the details, which are completely straightforward. ■

Claim 7. The identity (4) holds whenever $r = m/n$, where m and n are natural numbers (and c is any positive real).

Proof. We know that

$$\tan^m \left(\frac{\Pi(c)}{2} \right) = \tan \left(\frac{\Pi(mc)}{2} \right) \quad \text{(Claim 6)}$$

$$= \tan \left(\frac{\Pi \left(n \left(\frac{m}{n} c \right) \right)}{2} \right) \quad \text{(Algebra)}$$

$$= \tan^n \left(\frac{\Pi \left(\frac{m}{n} c \right)}{2} \right). \quad \text{(Claim 6: letting } \left(\frac{m}{n} c \right) \text{ play the role of } c)$$

Taking nth roots of both sides of this equation yields (4) when $r = m/n$. ■

---

[10] Retaining the parameter k, these become $\tan\Pi(x) = 1/\sinh(x/k)$, $\sin\Pi(x) = 1/\cosh(x/k)$, and $\cos\Pi(x) = \tanh(x/k)$.

Corollary. *The identity*

$$\tan^r\left(\frac{\Pi(c)}{2}\right) = \tan\left(\frac{\Pi(rc)}{2}\right) \tag{4}$$

holds for any positive numbers c and r.

Proof. For any fixed value of c, the two sides of the equation are continuous functions of r that agree on the positive rationals. Since the positive rationals are dense in the positive reals, the functions must agree on the positive reals. That is, the identity holds for any positive real numbers c and r. ■

With the identity (4) established for all positive r and c, Lobachevski sets $r = x/c$ (where x can be any positive real number), and thereby obtains that for any positive x,

$$\tan^{\frac{x}{c}}\left(\frac{\Pi(c)}{2}\right) = \tan\left(\frac{\Pi(x)}{2}\right), \tag{5}$$

He then selects his unit of measurement to be that which endows the expression $\tan(\Pi(1)/2)$ with the numerical value e^{-1}. This choice of unit implies that

$$\tan^{\frac{1}{c}}\left(\frac{\Pi(c)}{2}\right) = e^{-1},$$

which we can see by setting $x = 1$ in (5). Consequently, (5) becomes

$$\tan\left(\frac{\Pi(x)}{2}\right) = \tan^{\frac{x}{c}}\left(\frac{\Pi(c)}{2}\right) = \left(e^{-1}\right)^x = e^{-x},$$

which is the fundamental formula. ◆

Theory of Parallels 37

In this final proposition, Lobachevski develops the trigonometric formulae of the imaginary plane. Although we can now replace the Π-functions in any trigonometric equation with hyperbolic functions, Lobachevski chooses not to make these helpful translations. This lack, coupled with some awkward derivations and peculiar notation, make his work in this section appear particularly opaque. This is unfortunate, since the conclusions of this section are actually quite simple and admit easy proofs. To emphasize this fact, I shall derive Lobachevski's results in Π-free notation and deviate from his unnecessarily convoluted proofs of the two laws of cosines.

The Need for a New Rectilinear Relation

We can interpret the five gems as statements about right spherical triangles or right rectilinear triangles. In their spherical interpretation (see TP 35 notes), they specify the following trigonometric relationships:

A relation among the triangle's three sides.	(SG 5)
A relation among the two acute angles and a leg.	(SG 3,4)
A relation among the hypotenuse, a leg, **and the acute angle they *do not* include**.	(SG 1,2)

In the notes to TP 35, we developed all of spherical trigonometry from these relationships. Naturally, as soon as we have equations that specify the same three relations for right *rectilinear* triangles, we will be able to develop all of plane trigonometry by an analogous procedure.

The five gems, in their rectilinear interpretation, fall just short of providing these three relationships. In particular, they specify the following (see TP 36 notes):

A relation among the triangle's three sides.	(RG 1)
A relation among the two acute angles and a leg.	(RG 2,3)
A relation among the hypotenuse, a leg, **and their *included* angle**.	(RG 4,5)

If we can use these to derive a relation among a right rectilinear triangle's hypotenuse, leg, and the acute angle they do not include, then our trigonometric toolbox will be sufficiently powerful to develop all of trigonometry in the plane. It is easy to derive such a relation, but Lobachevski's words make the process seem mysterious. The required relation is the one that he calls (1) in the passage below.

The Need Satisfied

Of the five equations above (TP 35), the following two

$$\sin \Pi(c) = \sin \Pi(a) \sin \Pi(b)$$
$$\sin \Pi(\alpha) = \cos \Pi(\beta) \sin \Pi(b)$$

suffice to generate the other three: we can obtain one of the others by applying the second equation to side a rather than side b; we then deduce another by combining the equations already established. There will be no ambiguities of algebraic sign, since all angles here are acute.

Similarly, we obtain the two equations:

$$\tan \Pi(c) = \sin \Pi(\alpha) \tan \Pi(a) \tag{1}$$
$$\cos \Pi(a) = \cos \Pi(c) \cos \Pi(\beta). \tag{2}$$

The crucial relationship in this passage is (1). Using our translations from TP 36, we can rewrite this relationship in the form

$$\sin(A) = \frac{\sinh(a)}{\sinh(c)}.^{1} \tag{1'}$$

(Compare this with the familiar formula from plane Euclidean geometry: $\sin(A) = a/c$.)

Claim 1. In the generic right rectilinear triangle $[a, b, c; A, B]$, equation $(1')$ holds.

Proof. Since

$$\begin{aligned}
\sin^2(A) &= 1 - \cos^2(A) \\
&= 1 - \cosh^2(a) \sin^2(B) &&\text{(by RG 2 — see TP 36 notes)} \\
&= 1 - \cosh^2(a)[1 - \cos^2(B)] \\
&= 1 - \cosh^2(a)[1 - \cosh^2(b) \sin^2(A)] &&\text{(by RG 3)} \\
&= 1 - \cosh^2(a) + \cosh^2(a) \cosh^2(b) \sin^2(A) \\
&= 1 - \cosh^2(a) + \cosh^2(c) \sin^2(A) &&\text{(by RG 1)} \\
&= -\sinh^2(a) + \cosh^2(c) \sin^2(A),
\end{aligned}$$

a little algebra reveals that

$$\sinh^2(a) = [\cosh^2(c) - 1] \sin^2(A).$$

That is,

$$\sinh^2(a) = \sinh^2(c) \sin^2(A),$$

which implies equation $(1')$. ∎

---

[1] If we retain the parameter k, this equation becomes $\sin(A) = \sinh(a/k)/\sinh(c/k)$.

We now have our relation among a right triangle's hypotenuse, leg, and the non-included acute angle. Lobachevski will now use this relation to derive the law of sines in the plane, in precisely the way that we used its spherical analogue (SG1) to derive the law of sines on the sphere.

The Law of Sines for the Imaginary Plane

> "...with the mind I myself serve the law of God; but
> with the flesh the law of sin."
>
> —Romans 7:25

We shall now consider an arbitrary rectilinear triangle with sides a, b, c and opposite angles A, B, C.

If A and B are acute angles, then the perpendicular p dropped from C will fall within the triangle and cut side c into two parts: x, on the side of A, and $c - x$, on the side of B. This produces two right triangles. Applying equation (1) to each yields

$$\tan \Pi(a) = \sin(B) \tan \Pi(p).$$
$$\tan \Pi(b) = \sin(A) \tan \Pi(p).$$

These equations hold even if one of the angles, say B, is right or obtuse. Thus, for any rectilinear triangle whatsoever, we have

$$\sin(A) \tan \Pi(a) = \sin(B) \tan \Pi(b). \tag{3}$$

Equation (3) is the law of sines for the imaginary plane. Translated into Π-free notation and expressed in a more general form, it becomes

$$\frac{\sin(A)}{\sinh(a)} = \frac{\sin(B)}{\sinh(b)} = \frac{\sin(C)}{\sinh(c)}. \tag{3'}$$

As on the sphere or in the Euclidean plane, we can prove the law of sines in this context by dropping a perpendicular and working with the two resulting right triangles. This is precisely what Lobachevski does in his proof above. Here is the same proof, re-expressed in terms of hyperbolic functions.

Theorem 1. (Law of Sines) *In any rectilinear triangle $\triangle ABC$ in the imaginary plane with the usual labeling,[2]*

$$\frac{\sin(A)}{\sinh(a)} = \frac{\sin(B)}{\sinh(b)} = \frac{\sin(C)}{\sinh(c)}. \text{[3]}$$

---

[2] i.e. angles A, B, C lie opposite sides a, b, c.

[3] More generally, $\sin(A)/\sinh(a/k) = \sin(B)/\sinh(b/k) = \sin(C)/\sinh(c/k)$, for the positive parameter k.

Proof. Drop an altitude from C, and label the parts of the triangle as in the figures. Regardless of where the foot D of the perpendicular falls, we apply (1′) to the two right triangles that result, obtaining

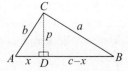

$$\sin(A) = \sinh(p)/\sinh(b),$$

and $\quad \sin(B) = \sinh(p)/\sinh(a).$[4]

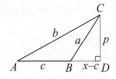

Solving both equations for $\sinh(p)$ and equating the results, we find that $\sin(A)/\sinh(a) = \sin(B)/\sinh(b)$. Repeating the argument, but dropping the perpendicular from A, yields $\sin(C)/\sinh(c) = \sin(B)/\sinh(b)$. Equating the two expressions for $\sin(B)/\sinh(b)$, we obtain the law of sines. ■

The First Law of Cosines for the Imaginary Plane

Lobachevski's derivation of the law of cosines is long-winded and his exposition of it is somewhat clumsy, but when we clear away the brambles, we find that his proof conforms to the expected mold: he drops an altitude to obtain two right triangles, applies known trigonometric formulae to them, and combines the results to produce the law of cosines. Breaking his convoluted proof into pieces and commenting on each part would only serve to reinforce the misconception that the law of cosines is difficult to establish. Thus, after presenting his argument whole, I shall follow it with an alternate, much cleaner derivation, which is essentially identical to the proof of the spherical law of cosines, or for that matter, the proof of the Euclidean law of cosines that appears in most high-school textbooks.

Applying equation (2) to a triangle with acute angles at A and B yields

$$\cos \Pi(x) = \cos(A) \cos \Pi(b)$$

$$\cos \Pi(c - x) = \cos(B) \cos \Pi(a).$$

These equations hold even when one of the angles A or B is right or obtuse.

For instance, when $B = \pi/2$, we have $x = c$; in this case, the first equation reduces to equation (2) and the second is trivially true. When $B > \pi/2$, applying equation (2) still yields the first equation; in place of the second, it yields $\cos \Pi(x - c) = \cos(\pi - B) \cos \Pi(a)$, which, however, is identical to the second, since $\cos \Pi(x - c) = -\cos \Pi(c - x)$ (TP 23), and $\cos(\pi - B) = -\cos(B)$. Finally, if A is right or obtuse, we must use $c - x$ and x, instead of x and $c - x$, so that the two equations will hold in this case also.

To eliminate x from the two equations above, we observe that

$$\cos \Pi(c - x) = \frac{1 - \left[\tan\left(\frac{\Pi(c-x)}{2}\right)\right]^2}{1 + \left[\tan\left(\frac{\Pi(c-x)}{2}\right)\right]^2}$$

$$= \frac{1 - e^{2x-2c}}{1 + e^{2x-2c}}$$

---

[4] If D happens to fall directly upon A or B, these two equations are trivially true. For example, if $D = A$, then $p = b$, and $\sin(A) = 1$, so the fact that $\sin(p) = \sin(b)\sin(A)$ is obviously true.

$$= \frac{1 - \left[\tan\left(\frac{\Pi(c)}{2}\right)\right]^2 \left[\cot\left(\frac{\Pi(x)}{2}\right)\right]^2}{1 + \left[\tan\left(\frac{\Pi(c)}{2}\right)\right]^2 \left[\cot\left(\frac{\Pi(x)}{2}\right)\right]^2}$$

$$= \frac{\cos\Pi(c) - \cos\Pi(x)}{1 - \cos\Pi(c)\cos\Pi(x)}.$$

If we substitute the expressions for $\cos\Pi(x)$ and $\cos\Pi(c - x)$ into this, it becomes

$$\cos\Pi(a)\cos(B) = \frac{\cos\Pi(c) - \cos(A)\cos\Pi(b)}{1 - \cos(A)\cos\Pi(b)\cos\Pi(c)},$$

from which it follows that

$$\cos\Pi(c) = \frac{\cos\Pi(a)\cos(B) + \cos\Pi(b)\cos(A)}{1 + \cos\Pi(a)\cos\Pi(b)\cos(A)\cos(B)}, \quad 5$$

and finally,

$$[\sin\Pi(c)]^2 = [1 - \cos(B)\cos\Pi(c)\cos\Pi(a)][1 - \cos(A)\cos\Pi(b)\cos\Pi(c)]. \quad (4)$$

Similarly, we also have

$$[\sin\Pi(a)]^2 = [1 - \cos(C)\cos\Pi(a)\cos\Pi(b)][1 - \cos(B)\cos\Pi(c)\cos\Pi(a)]$$

$$[\sin\Pi(b)]^2 = [1 - \cos(A)\cos\Pi(b)\cos\Pi(c)][1 - \cos(C)\cos\Pi(a)\cos\Pi(b)].$$

From these three equations, we find that

$$\frac{[\sin\Pi(b]^2 [\sin\Pi(c)]^2}{[\sin\Pi(a)]^2} = [1 - \cos(A)\cos\Pi(b)\cos\Pi(c)]^2.$$

From this it follows, without ambiguity of sign, that

$$\cos(A)\cos\Pi(b)\cos\Pi(c) + \frac{\sin\Pi(b)\sin\Pi(c)}{\sin\Pi(a)} = 1. \quad (5)$$

Such is Lobachevski's proof for the law of cosines (5). Here is a simpler approach, in which I have rewritten the law in Π-free notation.

Theorem 2. (First Law of Cosines) *In any rectilinear triangle $\triangle ABC$ in the imaginary plane with the usual labeling,*

$$\cosh(c) = \cosh(a)\cosh(b) - \sinh(a)\sinh(b)\cos(C).^6$$

Proof. If $\triangle ABC$ happens to be a right triangle, label it so that the right angle is at C. In this case, $\cos(C) = 0$, so the first law of cosines reduces to $\cos(c) = \cos(a)\cos(b)$. This is the

---

[5] In Lobachevski's original, the positions of this equation and the preceding one are reversed: presumably this was a printer's error. The fact that Halsted perpetuated it in his 1891 translation leads me to suspect that Halsted, realizing that one could reach the conclusions of TP 37 by simpler arguments than Lobachevski's own, did not bother to look very closely at the details as they stand.

[6] More generally, $\cosh(c/k) = \cosh(a/k)\cosh(b/k) - \sinh(a/k)\sinh(b/k)\cos(C)$, for the positive parameter k.

Pythagorean Theorem for the imaginary plane, which we have already proved in TP 36. (It is RG1.)

If $\triangle ABC$ is not a right triangle, we drop a perpendicular BD from B to AC. We will require slightly different proofs (different in details, but the same in spirit) for the case in which D falls inside AC, and the case in which D falls outside AC.

Case 1. (D lies within AC). As in the figure, we let $d = BD$, $p = AD$, and $q = DC$.

Then,

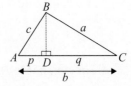

$$\begin{aligned}
\cosh(c) &= \cosh(d)\cosh(p) \quad \text{(RG1 on } \triangle ABD) \\
&= [\cosh(a)/\cosh(q)]\cosh(p) \quad \text{(RG1 on } \triangle BDC) \\
&= [\cosh(a)/\cosh(q)]\cosh(b-q) \\
&= [\cosh(a)/\cosh(q)][\cosh(b)\cosh(q) - \sinh(b)\sinh(q)] \\
&= \cosh(a)\cosh(b) - \sinh(b)\cosh(a)\tanh(q) \\
&= \cosh(a)\cosh(b) - \cosh(a)\sinh(b)[\tanh(a)\cos(C)] \quad \text{(RG3 on } \triangle BDC) \\
&= \cosh(a)\cosh(b) - \sinh(a)\sinh(b)\cos(C), \quad \text{as claimed.}
\end{aligned}$$

Case 2. (D falls outside of AC). Label the parts of the triangle as shown in the figure.

Then,

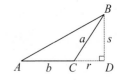

$$\begin{aligned}
\cosh(c) &= \cosh(b+r)\cosh(s) \quad \text{(RG1 on } \triangle ABD) \\
&= [\cosh(b)\cosh(r) + \sinh(b)\sinh(r)]\cosh(s) \quad \text{(trig identity)} \\
&= [\cosh(b)\cosh(r) + \sinh(b)\sinh(r)][\cosh(a)/\cosh(r)] \quad \text{(RG1 on } \triangle BCD) \\
&= \cosh(a)\cosh(b) + \sinh(b)\cosh(a)\tanh(r) \\
&= \cosh(a)\cosh(b) + \sinh(b)\cosh(a)[\cos(\pi - C)\tanh(a)] \quad \text{(RG4 on } \triangle BCD) \\
&= \cosh(a)\cosh(b) + \sinh(a)\sinh(b)\cos(\pi - C) \\
&= \cosh(a)\cosh(b) + \sin(a)\sin(b)\cos(C), \quad \text{as claimed.}
\end{aligned}$$

Thus, the first law of cosines holds for all rectilinear triangles in imaginary geometry. ∎

The Second Law of Cosines for the Imaginary Plane

Like spherical geometry, plane imaginary geometry admits AAA-congruence, and therefore admits a second law of cosines for determining a side when given all three angles. On the sphere, we obtained the second cosine law from the first by calling on the theory of polar triangles. Unfortunately, that is not an option in the plane. Lobachevski's proof is unnecessarily tortuous, so once again, I shall follow it with a simpler argument rather than detailing its twists and turns point by point.

The following expression for $\sin\Pi(c)$ follows from an alternate form of (3):

$$\sin\Pi(c) = \frac{\sin(A)}{\sin(C)}\tan\Pi(a)\cos\Pi(c).$$

If we substitute this expression into equation (5), we obtain

$$\cos\Pi(c) = \frac{\cos\Pi(a)\sin(c)}{\sin(A)\sin\Pi(b) + \cos(A)\sin(C)\cos\Pi(a)\cos\Pi(b)}.$$

If we substitute this expression for $\cos\Pi(c)$ into equation (4), we obtain

$$\cot(A)\sin(C)\sin\Pi(b) + \cos(C) = \frac{\cos\Pi(b)}{\cos\Pi(a)}. \tag{6}$$

By eliminating $\sin\Pi(b)$ with the help of equation (3), we find that

$$\frac{\cos\Pi(a)}{\cos\Pi(b)}\cos(C) = 1 - \frac{\cos(A)}{\sin(B)}\sin(C)\sin\Pi(a).$$

On the other hand, permuting the letters in equation (6) yields

$$\frac{\cos\Pi(a)}{\cos\Pi(b)} = \cot(B)\sin(C)\sin\Pi(a) + \cos(C).$$

By combining the last two equations, we obtain

$$\cos(A) + \cos(B)\cos(C) = \frac{\sin(B)\sin(C)}{\sin\Pi(a)}. \tag{7}$$

Here is a simpler approach to the second law of cosines, which I have rewritten in Π-free notation.

Theorem 3. (Second Law of Cosines) In any rectilinear triangle $\triangle ABC$ in the imaginary plane with the usual labeling,

$$\cos(C) = \sin(A)\sin(B)\cosh(c) - \cos(A)\cos(B).^{[7]}$$

Proof. If $\triangle ABC$ happens to be a right triangle, label it so that the right angle is at C. In this case, $\cos(C) = 0$, so the second law of cosines reduces to

$$\cos(A)\cos(B)/\cosh(c) = \sin(A)\sin(B).$$

By the Pythagorean theorem (RG1), this is equivalent to

$$\cos(A)\cos(B)/\cosh(a)\cosh(b) = \sin(A)\sin(B).$$

By RG 2 & 3, this is equivalent to $\sin(A)\sin(B) = \sin(A)\sin(B)$, or $1 = 1$. Thus, for any right triangle, the second law of cosines is equivalent to a trivially true statement.

If $\triangle ABC$ is not a right triangle, we drop a perpendicular BD from B to AC. We will require slightly different proofs (different in details, but the same in spirit) for the case in which D falls inside AC, and the case in which D falls outside AC.

---

[7] Or, if we retain the parameter $k > 0$, this becomes $\cos(C) = \sin(A)\sin(B)\cosh(c/k) - \cos(A)\cos(B)$.

Case 1. (*D* falls within *AC*). With the labels indicated in the figure, we have

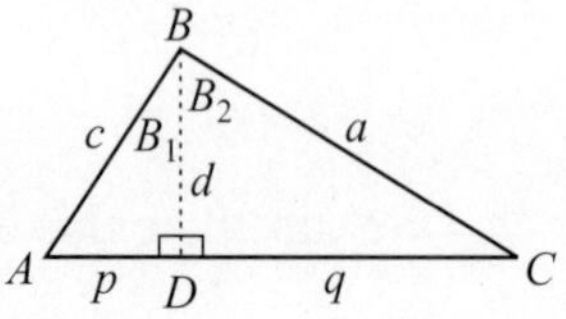

$$
\begin{aligned}
\cos(C) &= \cosh(d)\sin(B_2) \quad \text{(RG2 on } \triangle BDC) \\
&= \cosh(d)\sin(B - B_1) \\
&= \cosh(d)[\sin(B)\cos(B_1) - \sin(B_1)\cos(B)] \\
&= \cosh(d)\sin(B)\cos(B_1) - \cosh(d)\sin(B_1)\cos(B) \\
&= \cosh(d)\sin(B)\cosh(p)\sin(A) - \cos(A)\cos(B) \quad \text{(RG2 on} \triangle BDA) \\
&= \cosh(d)\sin(B)[\cosh(c)/\cosh(d)]\sin(A) - \cos(A)\cos(B) \quad \text{(RG1 on } \triangle BDA) \\
&= \sin(A)\sin(B)\cosh(c) - \cos(A)\cos(B).
\end{aligned}
$$

Case 2. (*D* falls outside of *AC*). Label the parts of the triangle as shown in the figure, where $\beta = \angle ABD$. Then,

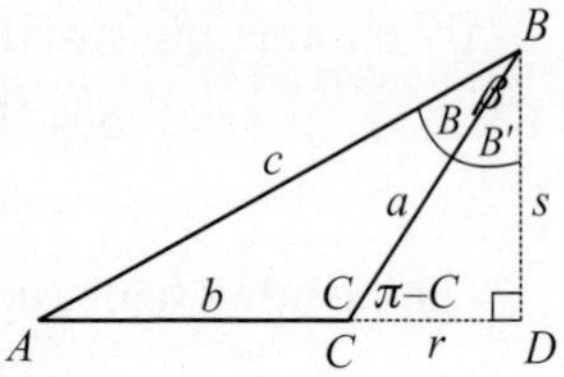

$$
\begin{aligned}
\cos(C) &= -\cos(\pi - C) \\
&= -\cosh(s)\sin(B') \quad \text{(RG2 on } \triangle CDB) \\
&= -\cosh(s)\sin(\beta - B) \\
&= -\cosh(s)[\sin(\beta)\cos(B) - \cos(\beta)\sin(B)] \\
&= \sin(B)\cos(\beta)\cosh(s) - \sin(\beta)\cosh(s)\cos(B) \\
&= \sin(B)\cos(\beta)\cosh(s) - \cos(A)\cos(B) \quad \text{(RG2 on } \triangle ADB) \\
&= \sin(B)\cos(\beta)[\cosh(c)/\cosh(b + r)] - \cos(A)\cos(B) \quad \text{(RG1 on } \triangle ADB) \\
&= [\cos(\beta)/\cosh(b + r)]\sin(B)\cosh(c) - \cos(A)\cos(B) \\
&= \sin(A)\sin(B)\cosh(c) - \cos(A)\cos(B). \quad \text{(RG2 on } \triangle ADB)
\end{aligned}
$$

Thus, the first law of cosines holds for all rectilinear triangles in imaginary geometry. ∎

Recapitulation of Trigonometric Formulae

Thus, the four equations that describe how the sides *a*, *b*, *c* and angles *A*, *B*, *C* are interrelated in rectilinear triangles are [equations (3), (5), (6), (7)]:

$$
\begin{cases}
\sin(A)\tan \Pi(a) = \sin(B)\tan \Pi(b) \\[2mm]
\cos(A)\cos\Pi(b)\cos\Pi(c) + \dfrac{\sin \Pi(b)\sin \Pi(c)}{\sin \Pi(a)} = 1 \\[2mm]
\cot(A)\sin(C)\sin\Pi(b) + \cos(C) = \dfrac{\cos \Pi(b)}{\cos \Pi(a)} \\[2mm]
\cos(A) + \cos(B)\cos(C) = \dfrac{\sin(B)\sin(C)}{\sin \Pi(a)}
\end{cases}
\tag{8}
$$

The first, second, and fourth of the relations collected in (8) are, respectively, the law of sines, the first law of cosines, and the second law of cosines. To understand why the remaining, unnamed equation is in their company, consider the following.

Any trigonometric relation necessarily involves four data (sides or angles), any three of which determine the entire triangle up to congruence and thus, in particular, determine the fourth datum. There are four distinct types of sets of four data: three sides and an angle (first law of cosines), three angles and a side (second law of cosines), two sides and their opposite angles (law of sines), or two sides and two angles, only one of which lies opposite an involved side. The unnamed relation in (8) is of this last type, and is included only for the sake of completeness[8]. Its theoretical importance is minimal, since the laws of sines and cosines already suffice to solve any "solvable" triangle. Lobachevski derived this relation as an intermediate result in his proof of the second law of cosines. For the sake of completeness, I will give a proof of this relation here, expressed in Π-free notation.

Claim 2. In any rectilinear triangle $\triangle ABC$ in the imaginary plane with the usual labeling, the following holds:

$$\frac{\cot(A)\sin(C)}{\cosh(b)} + \cos(C) = \frac{\tanh(b)}{\tanh(a)}.$$

Proof. This relation is an algebraic consequence of the laws of sines and cosines. In particular, we know that

$$\sin(C)\sinh(a) = \sin(A)\sinh(c) \qquad \text{(law of sines)}$$
$$= \sin(A)\left(\frac{\cosh(b)\cosh(c) - \cosh(a)}{\sinh(b)\cos(A)}\right) \qquad \text{(1st law of cosines)}$$
$$= \tan(A)\left(\frac{\cosh(b)\cosh(c) - \cosh(a)}{\sinh(b)}\right).$$

Multiplying both sides by $\cot(A)/\sinh(a)\cosh(b)$ yields

$$\frac{\cot A \sin C}{\cosh(b)} = \frac{\cosh(b)\cosh(c) - \cosh(a)}{\sinh(a)\sinh(b)\cosh(b)}$$
$$= \frac{\cosh(b)\left(\cosh(a)\cosh(b) - \sinh(a)\sinh(b)\cos(C)\right) - \cosh(a)}{\sinh(a)\sinh(b)\cosh(b)} \qquad \text{(1st law of cosines)}$$
$$= \frac{\cosh(a)\cosh(b)}{\sinh(a)\sinh(b)} - \cos(C) - \frac{\cosh(a)}{\sinh(a)\sinh(b)\cosh(b)}. \qquad \text{(algebra)}$$

Adding $\cos(C)$ to both sides, and putting the resulting right-hand side a common denominator gives us

$$\frac{\cot(A)\sin(C)}{\cosh(b)} + \cos(C) = \frac{\cosh(a)\cosh^2(b) - \cosh(a)}{\sinh(a)\sinh(b)\cosh(b)}$$
$$= \frac{\cosh(a)\left(\cosh^2(b) - 1\right)}{\sinh(a)\sinh(b)\cosh(b)}$$
$$= \frac{\cosh(a)\sinh^2(b)}{\sinh(a)\sinh(b)\cosh(b)}$$
$$= \frac{\tanh(b)}{\tanh(a)}.$$

∎

---

[8] To derive an analogous relation in Euclidean plane trigonometry, begin with the Euclidean law of cosines ($c^2 = a^2 + b^2 - 2ab\cos C$), and use the substitution $c = a\sin C/\sin A$ (law of sines) to obtain an equation involving two sides (a, b) and two angles (A, C), only one of which (A) lies opposite one of the two sides.

To supplement Lobachevski's summary, here are the key trigonometric formulae for recti-linear triangles in imaginary geometry, expressed in Π-free notation.

Law of sines
$$\frac{\sin(A)}{\sinh(a)} = \frac{\sin(B)}{\sinh(b)} = \frac{\sin(C)}{\sinh(c)}$$

First law of cosines $\cosh(c) = \cosh(a)\cosh(b) - \sinh(a)\sinh(b)\cos(C)$

Second law of cosines $\cos(C) = \sin(A)\sin(B)\cosh(c) - \cos(A)\cos(B).$

Approximations

When the sides a, b, c of the triangle are very small, we may content ourselves with the following approximations (TP 36):

$$\cot \Pi(a) = a,$$
$$\sin \Pi(a) = 1 - \tfrac{1}{2}a^2,$$
$$\cos \Pi(a) = a,$$

where the same approximations hold for sides b and c also.

Lobachevski will soon use the formulae of trigonometry to investigate what imaginary geometry looks like on the "infinitesimal" scale. He begins this process by approximating $\cot \Pi(x)$, $\sin \Pi(x)$, and $\cos \Pi(x)$ with simple functions whose accuracies approach perfection as x approaches zero.

Claim 3. $\cot \Pi(x) = x$, for infinitesimal values of x.[9]

Proof. Our translations from TP 36 give $\cot \Pi(x) = \sinh(x)$. Developing the right-hand side as a Taylor series and dropping higher order terms (which become increasingly insignificant compared to the leading term, as $x \to 0$) yields

$$\cot \Pi(x) = \sinh(x) = x + \frac{x^3}{3!} + \frac{x^5}{5!} + \cdots \approx x,$$

for small values of x. Thus, $\cot\Pi(x) = x$ for infinitesimally small values of x, as claimed. ∎

Corollary. $\tan \Pi(x) = 1/x$, *for infinitesimally small values of x.*

Claim 4. $\sin \Pi(x) = 1 - (x^2/2)$, for infinitesimally small values of x.

Proof. We know that

$$\sin \Pi(x) = \frac{1}{\cosh x} = \frac{1}{1 + \frac{x^2}{2!} + \frac{x^4}{4!} + \cdots} \approx \frac{1}{1 + \frac{x^2}{2}}.$$

Developing this last expression as a geometric series, and dropping higher order terms yields

$$\sin \Pi(x) \approx \frac{1}{1 + \frac{x^2}{2}} = 1 - \frac{x^2}{2} + \frac{x^4}{4} - \frac{x^6}{8} + \cdots \approx 1 - \frac{x^2}{2},$$

for small x. Thus, $\sin\Pi(x) = 1 - (x^2/2)$ for infinitesimally small values of x, as claimed. ∎

---

[9] That is, $\cot \Pi(x)/x \to 1$ as $x \to 0$. All other references to infinitesimals below should be interpreted as shorthand for statements about limiting behavior.

Claim 5. $\cos \Pi(x) = x$ for infinitesimal values of x.

Proof. For infinitesimal values of x, we may use the expressions that we have found in Claims 13 & 14 to write

$$\cos \Pi(x) = \cot \Pi(x) \sin \Pi(x) = x \left(1 - \frac{x^2}{2} \right) = x - \frac{x^3}{2} \approx x.$$

That is, $\cos \Pi(x) = x$ for infinitesimal values of x, as claimed. ∎

Under the Microscope:
Infinitesimal Imaginary Geometry = Euclidean Geometry

By substituting these approximations into the four trigonometric relations that Lobachevski summarizes in (8), we will obtain *approximate* relations, which becoming increasingly accurate when applied to smaller and smaller triangles (i.e. triangles that fit in smaller and smaller discs). In fact, we can guarantee that the difference between the exact and approximate values will be arbitrarily small if we restrict our attention to sufficiently tiny triangles. To avoid repeating precise but verbose statements about limiting behavior, we shall use the evocative if vague shorthand of describing our approximations as exact relations on infinitesimal triangles (i.e. triangles that fit in a disc of infinitesimal diameter).

For such small triangles, the equations (8) become

$$b \sin A = a \sin B$$
$$a^2 = b^2 + c^2 - 2bc \cos(A)$$
$$a \sin(A + C) = b \sin A$$
$$\cos A + \cos(B + C) = 0.$$

The first two of these equations are used in ordinary geometry; the last two equations lead, with help from the first, to the conclusion

$$A + B + C = \pi.$$

Therefore, imaginary geometry passes over into ordinary geometry when the sides of a rectilinear triangle are very small.

We can establish the four trigonometric relations for infinitesimal triangles by substituting our approximations into (8). The first two relations show that at a sufficiently small scale, the law of sines and the first law of cosines for imaginary geometry is indistinguishable from their Euclidean counterparts.

Claim 6. The *Euclidean* law of sines holds for any infinitesimal triangle $\triangle ABC$ in imaginary geometry.

Proof. All triangles in imaginary geometry satisfy the imaginary law of sines: $\sin(A) \tan \Pi(a) = \sin(B) \tan \Pi(b)$. Because the triangle is *infinitesimal*, we may use the substitution $\tan \Pi(x) = 1/x$ from Claim 3 (corollary), and rewrite this in its infinitesimal version, $\sin(A)/a = \sin(B)/b$, which is the ordinary Euclidean law of sines.[10] ∎

Claim 7. The Euclidean law of cosines holds for any infinitesimal triangle $\triangle ABC$ in imaginary geometry.

Proof. For any triangle in imaginary geometry, the first law of cosines (the 2nd equation in (8)) holds. For an *infinitesimal* triangle, we may use the substitutions given by Claims 4 & 5 to rewrite this cosine law as

$$bc \cos(A) + \frac{\left(1 - \frac{b^2}{2}\right)\left(1 - \frac{c^2}{2}\right)}{\left(1 - \frac{a^2}{2}\right)} = 1.$$

Or, after an algebraic massage,

$$b^2 c^2 + 2a^2 = 2b^2 + 2c^2 - 4bc \cos(A) + 2a^2 bc \cos(A).$$

At a sufficiently small scale, we may neglect terms of order 3 or higher; accordingly, we drop $b^2 c^2$ (order 4) and $2a^2 bc \cos(A)$ (order 3). After dropping them and dividing both sides of the resulting equation by 2, we obtain the infinitesimal version of the first law of cosines,

$$a^2 = b^2 + c^2 - 2bc \cos(A),$$

which is the Euclidean law of cosines, as claimed. ∎

The two remaining trigonometric relations for infinitesimal triangles in imaginary geometry do not correspond to named relations of Euclidean trigonometry, but when combined, they yield a still greater prize: the angle sum of an infinitesimal triangle in imaginary geometry is π.[11]

Claim 8. For any infinitesimal triangle $\triangle ABC$ in imaginary geometry, $a \sin(A + C) = b \sin(A)$.

Proof. For *any* triangle in imaginary geometry, the 3rd equation in (8) holds. For an *infinitesimal* triangle, we may use the substitutions given by Claims 4 & 5 to rewrite this relation as

$$\cot(A) \sin(C) \left(1 - \frac{b^2}{2}\right) + \cos(C) = \frac{b}{a}.$$

At a sufficiently small scale, we may choose to neglect terms of order 2 or higher. Dropping $b^2/2$ and multiplying both sides of the resulting equation by $a \sin(A)$ yields $a \left(\cos(A)\sin(C) + \cos(C)\sin(A)\right) = b \sin(A)$. Finally, a trigonometric identity lets us rewrite this as $a \sin(A + C) = b \sin(A)$, which was to be shown. ∎

---

[10] If we wanted to avoid Lobachevski's Π-function altogether, we could have used the Π-free formulation of the imaginary law of sines, $\sin(A)/\sinh(a) = \sin(B)/\sinh(b)$. Since $\sinh(x) \approx x$ for small x, we immediately have that $\sin(A)/a = \sin(B)/b$ for infinitesimal triangles. The next few claims could be handled the same way, using $\sinh(x) \approx x$ and $\cosh(x) \approx 1 + (x^2/2)$.

[11] Again, this infinitesimal language is just shorthand; a more accurate statement would be that, in imaginary geometry, the angle sum of a triangle approaches π as the size of the triangle decreases. In fact, this result follows immediately from Gauss' observation that angle defect and area are directly proportional in imaginary geometry (see the notes to TP 33), but since Lobachevski himself never mentions this proportionality within *The Theory of Parallels*, he obviously cannot invoke that theorem here.

Claim 9. For any infinitesimal triangle $\triangle ABC$ in imaginary geometry, $\cos(A) + \cos(B + C) = 0$.

Proof. For *any* triangle in imaginary geometry, the second law of cosines (4th equation in (8)) holds. For an *infinitesimal* triangle, we may use the substitutions given by Claims 4 & 5 to rewrite the second cosine law as

$$\cos(A) + \cos(B)\cos(C) = \frac{\sin(B)\sin(C)}{\left(1 - \frac{a^2}{2}\right)}.$$

Dropping the higher order term $a^2/2$ and rearranging the resulting equation yields

$$\cos(A) + [\cos(B)\cos(C) - \sin(B)\sin(C)] = 0.$$

That is, $\cos(A) + \cos(B + C) = 0$, as claimed. ∎

Claim 10. In imaginary geometry, every infinitesimal triangle $\triangle ABC$ has angle sum π.

Proof. Since

$$\begin{aligned}
0 &= \sin(B)\cos(B) - \sin(B)\cos(B) \\
&= [b\sin(A)/a]\cos(B) - \sin(B)\cos(B) && \text{(by Claim 6: infinitesimal law of sines)} \\
&= [b\sin(A)/a]\cos(B) + \sin(B)\cos(A + C) && \text{(by Claim 9, with the letters permuted)} \\
&= \sin(A + C)\cos(B) + \sin(B)\cos(A + C) && \text{(by Claim 8)} \\
&= \sin((A + C) + B) && \text{(trigonometric identity)} \\
&= \sin(A + B + C),
\end{aligned}$$

it follows that $(A + B + C)$, the angle sum of $\triangle ABC$, must be an integral multiple of π. Since the angle sum is obviously positive, and cannot exceed π (by the Saccheri-Legendre Theorem: TP 19), it must be π, as claimed. ∎

Because Euclid's parallel postulate is equivalent to the statement that triangles have angle sum π (see end of TP 22), the preceding result indicates that imaginary geometry becomes Euclidean at the infinitesimal scale. In other words, if one restricts one's attention to smaller and smaller portions of the imaginary plane, the phenomena that one observes will look increasingly Euclidean. Hence, *Euclidean geometry is a limiting case of imaginary geometry.*

Denouement

In the scholarly journal of the University of Kazan, I have published several investigations into the measurements of curves, plane figures, surfaces, and solids, as well as the application of imaginary geometry to analysis.

Beneath the surface of Lobachevski's innocent suggestion that readers of *The Theory of Parallels* might seek out his earlier Russian papers lie several layers of frustration and pathos. Let us work our way backwards through some of them, beginning with an event (or more accurately, non-event) that still lay in the future when Lobachevski wrote these hopeful words.

In the sixteen years of life that remained to Lobachevski, the mathematical community of Europe was to ignore or misunderstand *The Theory of Parallels*; no one wanted *more* of his work.

Even if he had found some sympathetic readers in Germany, France, or elsewhere in Europe, it is unlikely that they would have been able to secure, let alone read, his Russian publications.[12] Thus, in order to find readers, he had to contend not only with the "howls of the Boeotians" that Gauss so feared, but also a language barrier.

The one Russian mathematician who had Europe's ear in 1840 was Mikhail Ostrogradski, who makes an illuminating contrast with Lobachevski. Ostrogradski did impressive work in a fashionable area of mathematics, of which he wrote exclusively in French. Lobachevski, neither fashionable nor obliging, resented the suggestion that a Russian intellectual must abandon his native tongue to reach a receptive audience. "The language of a people," he argued, "is the testimony of its education, a true indicator of the degree of its enlightenment . . . ".[13]

One wonders how enlightened Russian mathematicians actually were at the time. A Russian review of Lobachevski's first publication on imaginary geometry, *On the Principles of Geometry* (1829–30), suggests the howling Boeotians whom Gauss feared so much. The review accused Lobachevski of "simpleminded ignorance" and declared that a more appropriate title for his work would have been, "A Satire on Geometry".[14] Undeterred, Lobachevski continued to publish his work on non-Euclidean geometry solely in Russian: *Imaginary Geometry* (1835), *The Applications of Imaginary Geometry to Some Integrals* (1836), and *New Principles of Geometry, with a Complete Theory of Parallels* (1835–8). His desired audience failed to materialize. Eventually, he gave in and wrote accounts of his work in German and French. By the time Lobachevski wrote *The Theory of Parallels*, he had already spent over a decade trying to publicize his work, in vain. His doomed efforts would continue for another fifteen years, ending only with his death in 1856.

In his early Russian papers, Lobachevski did indeed solve some mensuration problems in imaginary geometry, computing, for example, the circumference and area of a circle and the surface areas and volumes of spheres and tetrahedra. He was then able to evaluate some intractable definite integrals by interpreting them as magnitudes in imaginary geometry. Although he does not address these issues in *The Theory of Parallels*, he published another account of them in his final work, *Pangéométrie* (1855).

The Geometry of the Universe

> "What Vesalius was to Galen, what Copernicus was to Ptolemy, that was Lobachevsky to Euclid. There is, indeed, a somewhat instructive parallel between the last two cases. . . . Each of them has brought about a revolution in scientific ideas so great that it can only be compared with that wrought by the other. . . . they are changes in the conception of the Cosmos."
>
> —William Kingdon Clifford

In and of themselves, the equations (8) already constitute sufficient grounds for believing that the imaginary geometry might be possible. As a result, we have no

---

[12] The exception is Gauss, whose love of languages rivaled his feeling for mathematics: he knew some Russian, and after reading *The Theory of Parallels*, he sought out and read Lobachevski's earlier Russian works.

[13] Vucinich, p. 478.

[14] Rosenfeld, p. 209. The review was anonymous, but evidence suggests that Ostrogradski may have been responsible for it, either as author or instigator.

means other than astronomical observations with which to judge the accuracy that follows from calculations in the ordinary geometry. Its accuracy is very far-reaching, as I have demonstrated in one of my investigations; for example, in all angles whose sides we are capable of measuring, the sum of the three angles does not differ from π by so much as a hundredth of a second.

When Saccheri entered the counterintuitive world of imaginary geometry in the early 18th century, he anticipated a contradiction at every turn. He never did find one, but, undermined by his own fervent desire to do so, he eventually deluded himself into thinking that he had.[15]

In contrast, when Lobachevski described the same world a century later, he was convinced that the new geometry was at least as solid as Euclid's geometry, if not more so: its accuracy as a description of the physical universe might even surpass Euclid's. His plane trigonometric formulae (8) support this conviction. Since these equations imply that Euclidean geometry is a limiting case of imaginary geometry, the fact that the world "looks Euclidean" need not indicate that it actually *is* Euclidean. If the universe were very large—so large that even our telescopes could perceive only an infinitesimal portion of it—then physical space would appear Euclidean to us, even if imaginary geometry actually governs it.

What might Saccheri have thought of such an argument? He might well have thought of his countryman Galileo, who, only a few decades before Saccheri's birth, argued that the Earth is in constant motion despite the fact that it appears to be at rest. Might this analogy have swayed him into sympathy with Lobachevski's views? Saccheri and his contemporaries comprehended the vast size of the solar system,[16] but lacked any conception of the distances to the "fixed stars." They would have agreed that Italy is infinitesimal compared to the universe's size, but might still have maintained that the Earth's *orbit* encompasses a non-infinitesimal, though still small, portion of the universe. Thus, Saccheri might have maintained that if imaginary geometry governed the physical universe in the large, then astronomical measurements should reveal this fact. They do not. *Ergo, Euclides vindicatus est.*

By the time that Lobachevski published *The Theory of Parallels*, astronomers had discovered a very different picture of the universe. In 1780, William Herschel discovered Uranus, the first new planet found since antiquity. His later discovery of binary stars, orbiting one another, disposed of the last vestige of the Ptolemaic system, the "fixed crystalline sphere" upon which the stars all lie. Finally, his observations and calculations led him to the first estimate of our galaxy's diameter: 9000 light years. This enormous distance is in fact less than a tenth of the true value, but it represented a colossal step forward in humanity's understanding of just how big the universe truly is. Objections that might have been reasonable for Saccheri had become insupportable in the light of Herschel's discoveries. The conception of an overwhelmingly vast universe was, naturally, manifest in Laplace's masterpiece *Méchanique Céleste* (1799–1825). In his first published account of imaginary geometry, Lobachevski refers to

---

[15] According to Prékopa (p. 25), Imre Tóth has proposed that Saccheri's dubious "contradiction" was inserted because he feared the Inquisition. This is a fascinating thesis, even if Saccheri does seem a bit late for the Inquisition, but since Tóth's book, *God and Geometry*, is written in Hungarian, I have no way of learning more about it. The language barrier strikes again.

[16] In 1672, Cassini had calculated that the Earth's distance to the sun was approximately 87 million miles. This was still short of the true value (of approximately 93 million miles), but it was a vast improvement over Ptolemy's estimate (approximately 4 million miles). Because Kepler's third law (1619) establishes the *relative* distances of all of the known planets to the sun, Casini's measurement also measured the distances from the sun of every other known planet.

"...the view expressed by Laplace, that all the stars we see and the very Milky Way belong to merely one isolated cluster of heavenly bodies, similar to those that we perceive as faint shimmering spots in Orion...."[17]

The lesson that Lobachevski draws from this is significant,

"Nature itself reveals distances to us compared with which even the distance from Earth to the fixed stars disappears to insignificance." [18]

The first reasonably accurate measurement of a stellar distance came in 1838, when Friedrich Bessel showed that the distance from Earth to 61 Cygni, a faint "nearby" star, is over 270,000 times the distance from Earth to the sun.

The universe had become, in men's minds, much larger during the century that separates Lobachevski from Saccheri: large enough to suggest that *all* of our measurements—even astronomical measurements—might amount to infinitesimal distances with respect to it. In such a world, it was reasonable to conjecture that nature's large-scale geometry might be non-Euclidean. Its Euclidean appearance might be an illusion born of the fact that we make all our measurements from an insignificant corner of the cosmos.

Once we acknowledge the possibility that the geometry of the universe might be non-Euclidean, it is natural to want to test this hypothesis. Accordingly, Lobachevski analyzed astronomical records in an attempt to detect angle defect in an enormous triangle whose vertices lay at the Earth, the Sun, and Sirius, the brightest star in the night sky. An erroneous measurement of Sirius's parallax compromised his calculation, but in any case, he concluded that the defect must be (significantly) less than a hundredth of a second. Since we would expect this much experimental error even if Euclidean geometry were known to hold, this was not evidence in favor of imaginary geometry. Nor, of course, was this evidence in favor of Euclidean geometry. Instead, this calculation simply indicated that if the geometry of the universe *is* imaginary, we are too tiny to discern the discrepancy from Euclidean geometry.[19]

From the Imaginary Plane to the Sphere in the Blink of an i

Finally, it is worth observing that the four equations (8) of plane geometry become valid formulae of spherical geometry if we substitute $a\sqrt{-1}, b\sqrt{-1}, c\sqrt{-1}$ for the sides a, b, c ; these substitutions will change

$$\sin \Pi(a) \quad \text{to} \quad \frac{1}{\cos a},$$

$$\cos \Pi(a) \quad \text{to} \quad \sqrt{-1} \tan a,$$

$$\tan \Pi(a) \quad \text{to} \quad \frac{1}{\sqrt{-1} \sin a},$$

---

[17] Lobatschefskij, *Zwei geometrische Abhandlungen*, p. 24.

[18] *ibid.*, p. 24.

[19] For more information on Lobachevski's views on the relation between his geometry and the physical world, see Daniels.

and similarly for sides b and c. Hence, these substitutions change equations (8) into the following:

$$\sin A \sin b = \sin B \sin a$$

$$\cos a = \cos b \cos c + \sin b \sin c \cos A$$

$$\cot A \sin C + \cos C \cos b = \sin b \cot a$$

$$\cos A = \cos a \sin B \sin C - \cos B \cos C.$$

Lobachevski's last words in *The Theory of Parallels* hint at a blood relationship between the imaginary plane and the sphere. If we multiply all side lengths by i, we can magically transform the formulae of imaginary trigonometry into the formulae of spherical trigonometry. This is easy to see if we write the imaginary laws of cosines and sines in terms of hyperbolic functions.

For example, the first *imaginary* law of cosines is

$$\cosh(c) = \cosh(a)\cosh(b) + \sinh(a)\sinh(b)\cos(C).$$

If we multiply all the side lengths that appear in this equation by i, it becomes

$$\cosh(ci) = \cosh(ai)\cosh(bi) + \sin(ai)\sin(bi)\cos(C).$$

Since $\cosh(xi) = \cos(x)$ and $\sinh(xi) = i\sin(x)$, we can rewrite this as

$$\cos(c) = \cos(a)\cos(b) - \sin(a)\sin(b)\cos(C),$$

which is the first *spherical* law of cosines.

The same operation clearly transforms the imaginary law of sines and second law of cosines into the analogous trigonometric formulae for the sphere. Note that this process is reversible: multiplying the sides of a general spherical triangle by i will transform the spherical trigonometric laws into the corresponding imaginary laws. (This works because $\cos(xi) = \cosh(x)$ and $\sin(xi) = -i\sin(x)$.)

The relation between imaginary and spherical trigonometry is perhaps most manifest if we retain their respective parameters k and r. When we express them in this more general form, we can transform one law into the other simply by multiplying the *parameter* by i. For example, with their parameters expressed, the two laws of sines appear as follows:

	Imaginary			Spherical		
Laws of Sines	$\dfrac{\sin(A)}{\sinh\left(\frac{a}{k}\right)}$	$= \dfrac{\sin(B)}{\sinh\left(\frac{b}{k}\right)}$	$= \dfrac{\sin(C)}{\sinh\left(\frac{c}{k}\right)}$	$\dfrac{\sin(A)}{\sin\left(\frac{a}{r}\right)}$	$= \dfrac{\sin(B)}{\sin\left(\frac{a}{r}\right)}$	$= \dfrac{\sin(C)}{\sin\left(\frac{a}{r}\right)},$

where k and r are positive parameters whose numerical values depend on the unit of measurement used in each context. It is clear that multiplying the parameter by i in either case will transform the law into its counterpart in the other geometry. (Recall that $i^{-1} = -i$.) This seems to verify Lambert's hunch that a non-Euclidean geometry could make sense on a "sphere of imaginary radius."

The Consistency of Imaginary Geometry

"Lobatchevsky and Bolyai had considered this problem but had not been able to settle it."   —Morris Kline[20]

"Finally, it is not true that Lobachevsky's works do not contain a proof of the consistency of his geometry. Objectively, they do. The reduction of hyperbolic trigonometry to spherical is no less of a consistency proof of Lobachevsky's plane geometry than the Beltrami model."   —V. Ya. Perminov[21]

"Lobačevskiĭ's arguments do not represent a finished proof of the consistency of his plane geometry."   —B.A. Rosenfeld[22]

"In order to show that his 'imaginary' geometry or 'pangeometry' is as consistent as Euclidean geometry, Lobatschewsky pointed out that it is all based on his formulae for a triangle, which lead to the familiar formulae for a spherical triangle when the sides a, b, c are replaced by ia, ib, ic. Any inconsistency in the new geometry could be 'translated' into an inconsistency in spherical geometry (which is part of Euclidean geometry). Thus, after two thousand years of doubt, the independence of Euclid's postulate V was finally established."   —H.S.M. Coxeter[23]

"Lobachevsky. . . pondered over the problem all his life, but could not find a conclusive solution; this fell to the lot of future generations."   —V. Kagan[24]

This selection of quotations should serve as a corrective to the notion that mathematicians always agree as to what constitutes a proof and what does not. (It is also notable for exhibiting four distinct transliterations of Lobachevski's name, none of which agrees with my own.)

Lobachevski was convinced that contradictions would never appear in imaginary geometry, but few others shared his faith until compelled to do so by Beltrami and Poincaré, whose models of the imaginary plane definitively established imaginary geometry's consistency. These models, created more than a decade after Lobachevski's death, demonstrated that if a contradiction were to arise in imaginary geometry, a corresponding contradiction would also arise in Euclidean geometry. Hence, unless one is prepared to doubt the consistency of *Euclidean* geometry (which no one does), one has no right to question the consistency of imaginary geometry. Accordingly, we say that imaginary geometry is at least as consistent as Euclidean geometry.

Proving the relative consistency of a set of axioms by showing that any contradiction arising from them implies a contradiction in another (presumed consistent) area of mathematics has become a standard technique. For example, one can use it to show that Euclidean geometry is at least as consistent as the real number system, which itself is at least as consistent as the rational number system, which, in turn, is at least as consistent as the system of the natural numbers. The first to pioneer this technique, even if the attempt was not quite successful, seems to have been Lobachevski himself. In his *New Principles of Geometry* (1835–8, in Russian), he wrote,

---

[20] Kline, p. 914.

[21] Perminov, p. 19.

[22] Rosenfeld, p. 228.

[23] Coxeter, *Non-Euclidean Geometry*, p. 10.

[24] Kagan, p. 60.

We have found equations which represent the dependence of the angles and sides of a triangle [i.e. the laws of cosines and sines]. When, finally, we have given general expressions for elements of lines, areas and volumes of solids, all else in the Geometry is a matter of analytics, where calculations must necessarily agree with each other, and we cannot discover anything new that is not included in these first equations, . . . Thus, if one now needs to assume that some contradiction will force us subsequently to refute principles that we accepted in this geometry, then such contradiction can only hide in the very equations [i.e. the trigonometric equations]. We note, however, that these equations become equations of spherical Trigonometry as soon as, instead of the sides a, b, and c, we put $a\sqrt{-1}$, $b\sqrt{-1}$, and $c\sqrt{-1}$. . . therefore, ordinary Geometry, Trigonometry and the new Geometry will always agree among themselves.[25]

In this passage, Lobachevski seems to glimpse the future of geometry. To begin with, he hints at a remarkable synthesis of geometry and analysis: once the basic trigonometric relations are known, one can find expressions for the line, area, and volume elements (ds, dA, dV), and from thence, he suggests, one can answer (in principle) any geometric question. Thirty years after Lobachevski wrote these words, Bernhard Riemann gave a celebrated lecture in which he described how all of geometry is implicit in the Pythagorean theorem (a particular trigonometric relationship) and how different geometries ultimately stem from distinct versions of this theorem. Here, Lobachevski argues that because his trigonometric equations encompass all of imaginary geometry, any contradiction in imaginary geometry must be implicit within them. Such a contradiction, he intimates, is impossible: we can "translate" every statement of imaginary trigonometry into a corresponding statement of spherical trigonometry, so a contradiction in one context implies a contradiction in the other. Because there are no contradictions in spherical trigonometry, there can be none in imaginary trigonometry, and therefore, there can be no contradictions in imaginary geometry. Q.E.D.

This is a remarkable strategy for securing the consistency of his geometry, but it has a few flaws in its execution. First, even if we accept his claim that the higher parts of imaginary geometry are all consequences of the laws of cosines and sines, we still must worry about the logical status of these trigonometric equations. Even if no contradictions can arise from them, the possibility remains that the lower parts of imaginary geometry (i.e. the chain of theorems leading up to the trigonometric equations) may harbor one. Lobachevski addresses this objection in *Géométrie Imaginaire* (1837), his first non-Russian publication. In its pages, he develops imaginary geometry backwards. In *The Theory of Parallels*, he begins with the neutral axioms, assumes a new axiom (the angle sum is always less than π), and proceeds to deduce the trigonometric equations. In *Géométrie Imaginaire*, however, he begins with the neutral axioms, assumes the trigonometric equations, and deduces that the angle sum is always less than π. Thus, in the presence of the neutral axioms, the existence of angle defect (the added "Lobachevskian axiom") is equivalent to the formulae of imaginary trigonometry. Consequently, Lobachevski would seem to be justified in his claim that a contradiction anywhere in the imaginary geometry, regardless of whether it lurks in the higher or lower regions, would manifest itself in his trigonometric equations.

Even if we grant this, a fatal problem remains in his "translation" mechanism between imaginary and spherical geometries. Suppose that we wish to translate the proof of a theorem in

---

[25] Rosenfeld, p. 223

imaginary geometry into its spherical analogue. We must translate each step of the proof in turn. Those that are algebraic consequences of the trigonometric equations will translate smoothly, but when a step in the proof calls upon the neutral axioms or the theorems derived from them, trouble may appear. Suppose, for example, that one step involves a triangle in the imaginary plane and applies the neutral axioms to its sides, which are, of course, three lines in the plane. For the translation to work, we would need to apply the same neutral axioms to the sides of a spherical triangle, which are three great circles on the sphere. The neutral axioms, however, do *not* hold for such "lines"! Hence, our translation mechanism may break down at this point. We might find an *ad hoc* method to circumvent any given problem, but we cannot declare the consistency of imaginary geometry to be fully established until we have a translation mechanism that is guaranteed to function perfectly in every instance.

Although Lobachevski's method falls short of being a full proof of consistency, it nonetheless provides compelling evidence that imaginary geometry is at least as consistent as Euclidean geometry. Incidentally, *The Theory of Parallels* does (implicitly) contain a proof of the converse statement: any contradiction in Euclidean geometry would also manifest itself as a contradiction in the geometry of the horosphere, which in turn is a part of imaginary space. Hence, Euclidean geometry is at least as consistent as imaginary geometry. Combined with the famous converse of Beltrami and Poincaré, this result tells us that the two geometries stand or fall together. Astonishing though this fact seemed to the first mathematicians who recognized it, it would have come as no surprise to Lobachevski, who saw the two geometries not as separate warring entities, but as aspects of the same *pangeometry*, of which Euclid had investigated only the special case in which the angle of parallelism is fixed at $\pi/2$. By imposing his parallel postulate, Euclid had set the rest of pangeometry off limits. Lobachevski simply posed himself the task of charting the unexplored regions.

Bibliography

Alighieri, Dante. *Paradiso* (tr. Hollander). Doubleday, 2007.

Augustine of Hippo. *City of God Against the Pagans*. vols. XVI–XVIII.35 (eds. Eva Matthews Sanford and William McAllen Green). Harvard University Press, 1965.

Beckett, Samuel. *Murphy*. Grove Press, 1957.

Beltrami, Eugenio. "Essay on the Interpretation of Noneuclidean Geometry" (transl. Stillwell). In John Stillwell, *Sources of Hyperbolic Geometry*. American Mathematical Society, 1996: pp. 7–34.

Biermann, Kurt. " 'Ich bin im erschüttert': Neuer Versuch zur Aufklärung von Wachters Tod." ["I am profoundly shaken": New attempt at clarifying Wachter's death.] *Gauss-Ges. Göttingen Mitt.*, No. 35 (1998), pp. 41–43.

Bolyai, János. *The Science Absolute of Space* (transl. G.B. Halsted). Appendix in Bonola.

Bonola, Roberto. *Non-Euclidean Geometry* (transl. H.S. Carslaw). Dover Publications, Inc., 1955.

Borges, Jorge Luis. *Collected Fictions* (tr. Hurley). Viking, 1998.

———. *Labyrinths: Selected Stories and Other Writings* (transl. Yates, Irby, *et al.*). New Directions Publishing Corporation, 1962.

Bressoud, David. *A Radical Approach to Real Analysis*. Mathematical Association of America, 1994.

Carroll, Lewis. *Euclid and his Modern Rivals*. Dover Publications, Inc., 1973.

Clifford, William Kingdon. *Lectures and Essays* (Vol. 1). MacMillan and Co., 1879.

Coolidge, Julian Lowell. *A History of Geometrical Methods*. Dover Publications, Inc., 1963.

Coxeter, H.S.M. *Introduction to Geometry*. John Wiley & Sons, Inc., 1961.

———. *Non-Euclidean Geometry* (4th ed.). University of Toronto Press, 1961.

Daniels, Norman. "Lobachevsky: Some Anticipations of Later Views on the Relation between Geometry and Physics". *Isis*, Vol. 66, No.1 (Mar., 1975), pp. 75–85.

Dictionary of Scientific Biography. 16 vols. Scribner. 1970-1990.

Dunnington, G.W. *Gauss: Titan of Science*. Mathematical Association of America, 2004.

Engel, F. and Stäckel, P. *Theorie der Parallellinien von Euklid bis auf Gauss*. Teubner, 1913.

Euclid. *The Thirteen Books of the Elements* (transl. Sir Thomas Heath). 3 vols. Dover Publications, Inc., 1956.

Eves, Howard. *A Survey of Geometry*. 2 vols. Allyn and Bacon, Inc. 1963.

Fauvel, J.G. and Gray, J.J. *The History of Mathematics: A Reader*. Macmillan, 1987.

Gauss, Karl Friedrich. *Werke*, VIII. Georg Olms Verlag, 1973.

Gray, Jeremy. *Ideas of Space: Euclidean, Non-Euclidean and Relativistic* (2nd ed.). Oxford University Press, 1989.

———. *János Bolyai, Non-Euclidean Geometry, and the Nature of Space*. Burndy Library Publications, 2004.

Greenberg, Marvin. *Euclidean and Non-Euclidean Geometries: Development and History* (2nd ed.). W.H. Freeman and Company, 1980.

Hartshorne, Robin. *Geometry: Euclid and Beyond*. Springer-Verlag New York, Inc., 2000.

Hilbert, David. *Foundations of Geometry* (transl. Leo Unger). Open Court Publishing Company, 1971.

Kagan, V. *N. Lobachevsky and his Contribution to Science*. Foreign Languages Publishing House, Moscow, 1957.

Kline, Morris. *Mathematical Thought from Ancient to Modern Times*. Oxford University Press, 1972.

Laubenbacher, Reinhard & Pengelley, David. *Mathematical Expeditions: Chronicles by the Explorers*. Springer-Verlag, 1998.

Lobachevski, Nicolai Ivanovich. *New Principles of Geometry with Complete Theory of Parallels* (transl. Halsted). The Neomon, 1897.

———. "Géométrie Imaginaire." *Journal für die reine und angewandte Mathematik*. Vol. 17 (1837), pp. 295–320.

Lobatcheffsky, N., *Pangéométrie, ou, Précis de géométrie fondée sur une théorie générale et rigoureuse des parallels*, Imprimerie de l'Université, Kasan, 1855.

Lobatschefskij, Nikolaj Iwanowitsch. *Zwei geometrische Abhandlungen* (transl. F. Engel). Teubner, 1898.

Lobatschewsky, Nicolaus. *Geometrische Untersuchungen zur Theorie der Parallellinien*. Fincke, 1840.

Minding, Ferdinand. "Beiträge zur Theorie der kürtzesten Linien auf krummen Flächen." *Journal für die reine und angewandte Mathematik*. Vol. 20 (1840), pp. 323–327.

Moise, E.E. *Elementary Geometry from an Advanced Standpoint* (2nd ed.). Addison-Wesley, 1974.

Pasch, Moritz. *Vorlesungen über neuere Geometrie*. Teubner, 1882.

The Norton Anthology of Poetry (ed. Allison, Barrows, *et al.*). W.W. Norton & Company, 1975.

Perminov, V. Ya. "The Philosophical and Methodological Thought of N.I. Lobachevsky." *Philosophia Mathematica* (3) Vol. 5 (1997), pp. 3–20.

Prékopa, András & Molnár, Emil (editors). *Non-Euclidean Geometries: János Bolyai Memorial Volume*. Springer-Verlag, 2005.

Rosenfeld, B.A. *A History of Non-Euclidean Geometry: Evolution of the Concept of a Geometric Space* (transl. Abe Shenitzer). Springer-Verlag New York, Inc., 1988.

Rothman, Tony. "Genius and Biographers: The Fictionalization of Evariste Galois," *American Mathematical Monthly*, Vol. 89 (1982), pp. 84–106.

Russell, Bertrand. *The Autobiography of Bertrand Russell: 1872–1914*. Little, Brown, and Co., 1967.

———. *Mysticism and Logic*. Doubleday & Company, Inc., 1957.

Saccheri, Gerolamo. *Euclides Vindicatus* (transl. George Bruce Halsted). Chelsea Publishing Co., 1986.

Sommerville, D.M.Y. *The Elements of Non-Euclidean Geometry*. Dover Publications, Inc., 1958.

Stahl, Saul. *The Poincaré Half-Plane: A Gateway to Modern Geometry*. Jones and Bartlett Publishers, Inc., 1993.

Sterne, Laurence. *The Life and Opinions of Tristram Shandy, Gentleman*. The Heritage Press, 1935.

Stillwell, John. *Sources of Hyperbolic Geometry*. American Mathematical Society, 1996.

Vucinich, Alexander. "Nikolai Ivanovich Lobachevskii: The Man Behind the First Non-Euclidean Geometry." *Isis*, Vol. 53, No. 4 (Dec., 1962), pp. 465–481.

Wolfe, Harold E.. *Introduction to Non-Euclidean Geometry*. Holt, Rinehart and Winston, Inc., 1945.

Appendix: Nicolai Ivanovich Lobachevski's Theory of Parallels

In geometry, I have identified several imperfections, which I hold responsible for the fact that this science, apart from its translation into analysis, has taken no step forward from the state in which it came to us from Euclid. I consider the following to be among these imperfections: vagueness in the basic notions of geometric magnitudes, obscurity in the method and manner of representing the measurements of such magnitudes, and finally, the crucial gap in the theory of parallels. Until now, all mathematicians' efforts to fill this gap have been fruitless. Legendre's labors in this area have contributed nothing. He was forced to abandon the one rigorous road, turn down a side path, and seek sanctuary in extraneous propositions, taking pains to present them—in fallacious arguments—as necessary axioms.

I published my first essay on the foundations of geometry in the "Kazan Messenger" in the year 1829. Hoping to provide an essentially complete theory, I then undertook an exposition of the subject in its entirety, publishing my work in installments in the "Scholarly Journal of the University of Kazan" in the years 1836, 1837, and 1838, under the title, "New Principles of Geometry, with a Complete Theory of Parallels". Perhaps it was the extent of this work that discouraged my countrymen from attending to its subject, which had ceased to be fashionable since Legendre. Be that as it may, I maintain that the theory of parallels should not forfeit its claim to the attentions of geometers. Therefore, I intend here to expound the essence of my investigations, noting in advance that, contrary to Legendre's opinion, all other imperfections, such as the definition of the straight line, will prove themselves quite foreign here and without any real influence on the theory of parallels. Lest my reader become fatigued by a multitude of theorems whose proofs present no difficulties, I shall list here in the preface only those that will actually be required later.

1) *A straight line covers itself in all its positions.* By this, I mean that a straight line will not change its position during a rotation of a plane containing it if the line passes through two fixed points in the plane.

2) *Two straight lines cannot intersect one another in two points.*

3) *By extending both sides of a straight line sufficiently far, it will break out of any bounded region. In particular, it will separate a bounded plane region into two parts.*

4) *Two straight lines perpendicular to a third will never intersect one another, regardless of how far they are extended.*

5) *When a straight lines passes from one side to the other of a second straight line, the lines always intersect.*

6) *Vertical angles, those for which the sides of one angle are the extensions of the other, are equal. This is true regardless of whether the vertical angles lie in the plane or on the surface of a sphere.*

7) *Two straight lines cannot intersect if a third line cuts them at equal angles.*

8) *In a rectilinear triangle, equal sides lie opposite equal angles, and conversely.*

9) *In rectilinear triangles, greater sides and angles lie opposite one another. In a right triangle, the hypotenuse is greater than either leg, and the two angles adjacent to it are acute.*

10) *Rectilinear triangles are congruent if they have a side and two angles equal, two sides and their included angle equal, two sides and the angle that lies opposite the greatest side equal, or three sides equal.*

11) *If a straight line is perpendicular to two intersecting lines, but does not lie in their common plane, then it is perpendicular to all straight lines in their common plane that pass through their point of intersection.*

12) *The intersection of a sphere with a plane is a circle.*

13) *If a straight line is perpendicular to the intersection of two perpendicular planes and lies in one of them, then it is perpendicular to the other plane.*

14) *In a spherical triangle, equal angles lie opposite equal sides, and conversely.*

15) *Spherical triangles are congruent if they have two sides and their included angle equal, or one side and its adjacent angles equal.*

Explanations and proofs shall accompany the theorems from now on.

Proposition 16

In a plane, all lines that emanate from a point can be partitioned into two classes with respect to a given line in the same plane; namely, those that cut the given line and those that do not cut it.

The boundary-line separating the classes from one another shall be called a *parallel* to the given line.

From point A (see figure at right), drop the perpendicular AD to the line BC, and erect the perpendicular AE upon it. Now, either all of the lines entering the right angle $\angle EAD$ through A will, like AF in the figure, cut DC, or some of these lines will not cut DC, resembling the perpendicular AE in this respect. The uncertainty as to whether the perpendicular AE is the only line that fails to cut DC requires us to suppose it possible

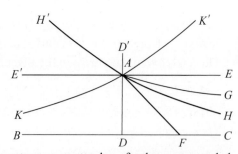

that there are still other lines, such as AG, which do not cut, no matter how far they are extended.

At the transition from the cutting lines such as AF to the non-cutting lines such as AG, one necessarily encounters a parallel to DC. That is, one will encounter a boundary line AH with the property that all the lines on one side of it, such as AG, do not cut DC, while all the lines on the other side of it, such as AF, do cut DC.

The angle $\angle HAD$ between the parallel AH and the perpendicular AD is called the *angle of parallelism*; we shall denote it here by $\Pi(p)$, where $p = AD$. If $\Pi(p)$ is a right angle, then the extension AE' of AE will be parallel to the extension DB of the line DC. Observing the four right angles formed at point A by the perpendiculars AE, AD, and their extensions AE' and AD', we note that any line emanating from A has the property that either it or its extension lies in one of the two right angles facing BC. Consequently, with the exception of the parallel EE', all lines through A will cut the line BC when sufficiently extended.

If $\Pi(p) < \pi/2$, then the line AK, which lies on the other side of AD and makes the same angle $\angle DAK = \Pi(p)$ with it, will be parallel to the extension DB of the line DC. Hence, under this hypothesis we must distinguish directions of parallelism.

Among the other lines that enter either of the two right angles facing BC, those lying between the parallels (i.e. those within the angle $\angle HAK = 2\Pi(p)$) belong to the class of cutting-lines. On the other hand, those that lie between either of the parallels and EE' (i.e. those within either of the two angles $\angle EAH = \pi/2 - \Pi(p)$ or $\angle E'AK = \pi/2 - \Pi(p)$) belong, like AG, to the class of non-cutting lines.

Similarly, on the other side of the line EE', the extensions AH' and AK' of AH and AK are parallel to BC; the others are cutting-lines if they lie in the angle $\angle K'AH'$, but are non-cutting lines if they lie in either of the angles $\angle K'AH'$ or $\angle H'AE'$.

Consequently, under the presupposition that $\Pi(p) = \pi/2$, lines can only be cutting-lines or parallels. However, if one assumes that $\Pi(p) < \pi/2$, then one must admit two parallels, one on each side. Furthermore, among the remaining lines, one must distinguish between those that cut and those that do not cut. Under either assumption, the distinguishing mark of parallelism is that the line becomes a cutting line when subjected to the smallest deviation toward the side where the parallel lies. Thus, if AH is parallel to DC, then regardless of how small the angle $\angle HAF$ may be, the line AF will cut DC.

Proposition 17

A straight line retains the distinguishing mark of parallelism at all its points.

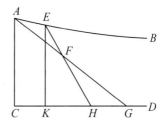

Let AB be parallel to CD, with AC perpendicular to the latter. We shall examine two points, one chosen arbitrarily from the line AB and one chosen arbitrarily from its extension beyond the perpendicular.

Let E be a point on that side of the perpendicular in which AB is parallel to BC. From E, drop a perpendicular EK to CD, and draw any line EF lying within the angle $\angle BEK$. Draw the line through the points A and F. Its extension must intersect CD (by TP 16) at some point G. This produces a triangle $\triangle ACG$, which is pierced by the line EF. This line, by construction, cannot intersect AC; nor can it intersect AG or EK a second time (TP 2). Hence, it must meet CD at some point H (by TP 3).

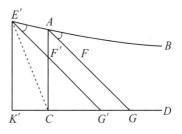

Now let E' be a point on the extension of AB, and drop a perpendicular $E'K'$ to the extension of the line CD. Draw any line $E'F'$ with the angle $\angle AE'F'$ small enough to cut AC at some point F'. At the same angle of inclination towards AB, draw a line AF; its extension will intersect CD

(by TP 16) at some point G. This construction produces a triangle $\triangle AGC$, which is pierced by the extension of line $E'F'$. This line can neither cut AC a second time, nor can it cut AG, since $\angle BAG = \angle BE'G'$ (by TP 7). Thus, it must meet CD at some point G'.

Therefore, regardless of which points E and E' the lines EF and $E'F'$ emanate from, and regardless of how little these lines deviate from AB, they will always cut CD, the line to which AB is parallel.

Proposition 18

Two parallel lines are always mutually parallel.

Let AC be perpendicular to CD, a line to which AB is parallel. From C, draw any line CE making an acute angle $\angle ECD$ with CD. From A, drop the perpendicular AF to CE. This produces a right triangle $\triangle ACF$, in which the hypotenuse AC is greater than the side AF (TP 9).

If we make $AG = AF$ and lay AF upon AG, the lines AB and FE will assume positions AK and GH in such a way that $\angle BAK = \angle FAC$. Consequently, AK must intersect the line DC at some point K (TP 16), giving rise to a triangle $\triangle AKC$. The perpendicular GH within this triangle must meet the line AK at some point L (TP 3). Measured along AB from A, the distance AL determines the intersection point of the lines AB and CE. Therefore, CE will always intersect AB, regardless of how small the angle $\angle ECD$ may be. Hence, CD is parallel to AB (TP 16).

Proposition 19

In a rectilinear triangle, the sum of the three angles cannot exceed two right angles.

Suppose that the sum of the three angles in a triangle is $\pi + \alpha$.

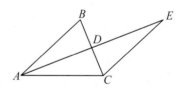

Bisect the smallest side BC at D, draw the line AD, make its extension DE equal to AD, and draw the straight line EC. In the congruent triangles $\triangle ADB$ and $\triangle CDE$ (TP 16 and TP 10), we have $\angle ABD = \angle DCE$ and $\angle BAD = \angle DEC$. From this, it follows that the sum of the three angles in $\triangle ACE$ must also be $\pi + \alpha$. We note additionally that $\angle BAC$, the smallest angle of $\triangle ABC$ (TP 9), has been split into two parts of the new triangle $\triangle ACE$; namely, the angles $\angle EAC$ and $\angle AEC$.

Continuing in this manner, always bisecting the side lying opposite the smallest angle, we eventually obtain a triangle in which $\pi + \alpha$ is the sum of the three angles, two of which are smaller than $\alpha/2$ in absolute magnitude. Since the third angle cannot exceed π, α must be either zero or negative.

Proposition 20

If the sum of the three angles in one rectilinear triangle is equal to two right angles, the same is true for every other triangle.

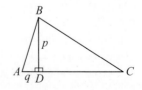

If we suppose that the sum of the three angles in triangle $\triangle ABC$ is equal to π, then at least two of its angles, A and C, must be acute.

From the third vertex, B, drop a perpendicular p to the opposite side, AC. This will split the triangle $\triangle ABC$ into two right triangles. In each of these, the angle sum will also be π: neither angle sum can exceed π (TP 19), and the fact that the right triangles comprise triangle $\triangle ABC$ ensures that neither angle sum is less than π.

In this way, we obtain a right triangle whose legs are p and q; from this we can obtain a quadrilateral whose opposite sides are equal, and whose adjacent sides are perpendicular. By repeated application of this quadrilateral, we can construct another with sides np and q, and eventually a quadrilateral $EFGH$, whose adjacent sides are perpendicular, and in which $EF = np$, $EH = mq$, $HG = np$, and $FG = mq$, where m and n can be any whole numbers. The diagonal FH of such a quadrilateral

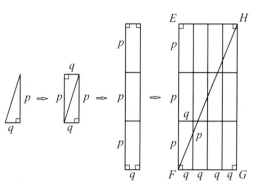

divides it into two congruent right triangles, $\triangle FEH$ and $\triangle FGH$, each of which has angle sum π.

The numbers m and n can always be chosen so large that any given right triangle $\triangle JKL$ can be enclosed within a right triangle $\triangle JMN$, whose arms are $NJ = np$ and $MJ = mq$, when one brings their right angles into coincidence. Drawing the line LM yields a sequence of right triangles in which each successive pair shares a common side.

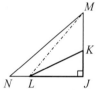

The triangle $\triangle JMN$ arises as the union of the triangles $\triangle NML$ and $\triangle JML$. The angle sum exceeds π in neither of these; it must, therefore, equal π in each case in order to make the composite triangle's angle sum equal to π. Similarly, the triangle $\triangle JML$ consists of the two triangles $\triangle KLM$ and $\triangle JKL$, from which it follows that the angle sum of $\triangle JKL$ must equal π.

In general, this must be true of every triangle since each triangle can be cut into two right triangles. Consequently, only two hypotheses are admissible: the sum of the three angles either equals π for all rectilinear triangles, or is less than π for all rectilinear triangles.

Proposition 21

From a given point, one can always draw a straight line that meets a given line at an arbitrarily small angle.

From the given point A, drop the perpendicular AB to the given line BC; choose an arbitrary point D on BC; draw the line AD; make $DE = AD$, and draw AE. If we let $\alpha = \angle ADB$ in the right triangle $\triangle ABD$, then the angle $\angle AED$ in the isosceles triangle $\triangle ADE$ must be less than or equal to $\alpha/2$ (TP 8

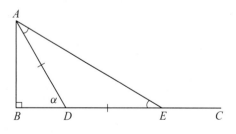

& 19)[1]. Continuing in this manner, one eventually obtains an angle $\angle AEB$ that is smaller than any given angle.

---

[1] I have corrected an apparent misprint occurring in Lobachevski's text and perpetuated in Halsted's 1891 translation of *TP*. In these sources, Lobachevski cites TP 20 at this point, rather than TP 19. This makes little sense; TP 20 relates the

Proposition 22

If two perpendiculars to the same straight line are parallel to one another, then the sum of the three angles in all rectilinear triangles is π.

Let the lines AB and CD be parallel to one another and perpendicular to AC. From A, draw lines AE and AF to points E and F chosen anywhere on the line CD such that $FC > EC$. If the sum of the three angles equals $\pi - \alpha$ in the right triangle $\triangle ACE$ and $\pi - \beta$ in triangle $\triangle AEF$, then it must equal $\pi - \alpha - \beta$ in triangle $\triangle ACF$, where α and β cannot be negative. Further, if we let $a = \angle BAF$ and $b = \angle AFC$, then $\alpha + \beta = a - b$.

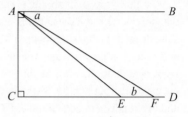

By rotating the line AF away from the perpendicular AC, one can make the angle a between AF and the parallel AB as small as one wishes; one reduces the angle b by the same means. It follows that the magnitudes of the angles α and β can be none other than $\alpha = 0$ and $\beta = 0$.

From what we have seen thus far, it follows either that the sum of the three angles in all rectilinear triangles is π, while the angle of parallelism $\Pi(p) = \pi/2$ for all lines p, or that the angle sum is less than π for all triangles, while $\Pi(p) < \pi/2$ for all lines p. The first hypothesis serves as the foundation of the ordinary geometry and plane trigonometry.

The second hypothesis can also be admitted without leading to a single contradiction, establishing a new geometric science, which I have named Imaginary Geometry, which I intend to expound here as far as the derivation of the equations relating the sides and angles of rectilinear and spherical triangles.

Proposition 23

For any given angle α, *there is a line* p *such that* $\Pi(p) = \alpha$.

Let AB and AC be two straight lines forming an acute angle α at their point of intersection A. From an arbitrary point B' on AB, drop a perpendicular $B'A'$ to AC. Make $A'A'' = AA'$, and erect a perpendicular $A''B''$ upon A''; repeat this construction until reaching a perpendicular CD that fails to meet AB. This must occur, for if the sum of the three angles equals π-a in triangle $\triangle AA'B'$, then it equals $\pi - 2a$ in triangle $\triangle AB'A''$, and is less than $\pi - 2a$ in $\triangle AA''B''$ (TP 20); if the construction could be repeated indefinitely, the sum would eventually become negative,

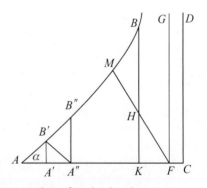

thereby demonstrating the impossibility of the perpetual construction of such triangles.

The perpendicular CD itself might have the property that all other perpendiculars closer to A cut AB. At any rate, there is a perpendicular FG at the transition from the cutting-perpendiculars to the non-cutting-perpendiculars that does have this property. Draw any line FH making an acute angle with FG and lying on the same side of it as point A. From any point H of FH, drop a perpendicular HK to AC; its extension must intersect AB at some point; say, at B. In this way,

---

angle sum of one triangle to the angle sums of all triangles – an issue having scarcely anything to do with the present proposition's modest concerns. - SB

the construction yields a triangle $\triangle AKB$, into which the line FH enters and must, consequently, meet the hypotenuse AB at some point M. Since the angle $\angle GFH$ is arbitrary and can be chosen as small as one wishes, FG is parallel to AB, and $AF = p$. (TP 16 and 18).

It is easy to see that with the decrease of p, the angle α increases, approaching the value $\pi/2$ for $p = 0$; with the increase of p, the angle α decreases, approaching ever nearer to zero for $p = \infty$.

Since we are completely free to choose the angle that shall be assigned to the symbol $\Pi(p)$ when p is a negative number, we shall adopt the convention that $\Pi(p) + \Pi(-p) = \pi$, an equation which gives the symbol a meaning for all values of p, positive as well as negative, and for $p = 0$.

Proposition 24

The farther parallel lines are extended in the direction of their parallelism, the more they approach one another.

Upon the line AB, erect two perpendiculars $AC = BD$, and join their endpoints C and D with a straight line. The resulting quadrilateral $CABD$ will have right angles at A and B, but acute angles at C and D (TP 22[2]). These acute angles are equal to one another; one can easily convince oneself of this by imagining laying the quadrilateral upon itself in such a way that the line BD lies upon AC, and AC lies upon BD. Bisect AB. From the midpoint E, erect the line EF perpendicular to AB; it will be perpendicular to CD as well, since the quadrilaterals $CAEF$ and $FEBD$ coincide when one is laid on top of the other in such a way that FE remains in the same place.

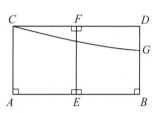

Consequently, the line CD cannot be parallel with AB. On the contrary, the line from point C that *is* parallel to AB, which we shall call CG, must incline toward AB (TP 16), cutting from the perpendicular BD a part $BG < CA$. Since C is an arbitrary point of the line CG, it follows that the farther CG is extended, the nearer it approaches AB.

Proposition 25

Two straight lines parallel to a third line are parallel to one another.

We shall first assume that the three lines AB, CD, and EF lie in one plane.

If one of the outer lines, say AB, and the middle line, CD, are parallel to the remaining outer line, EF, then AB and CD will be parallel to one another. To prove this, drop a perpendicular AE from any point A of AB to EF; it will intersect CD at some point C (TP 5), and the angle $\angle DCE$ will be acute (TP 22). Drop a perpendicular AG from A to CD; its foot G must fall on the side of C that forms an acute angle with AC (TP 9). Every line AH drawn from A into angle $\angle BAC$ must cut EF, the parallel to AB, at some point H, regardless of how small the angle $\angle BAH$ is taken. Consequently, the line CD, which enters the triangle $\triangle AEH$, must cut the line AH at

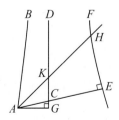

---

[2] This refers to Lobachevski's declaration at the end TP 22 that he would work in imaginary geometry from that point forward. Had he carried out this construction earlier, he would not have been able to deduce that the angles at C and D were acute; in neutral geometry, they could be either acute or right. - SB

some point K, since it is impossible for it to leave the triangle through EH. When AH is drawn from A into the angle $\angle CAG$, it must cut the extension of CD between C and G in the triangle $\triangle CAG$. From the preceding argument, it follows that AB and CD are parallel (TP 16 and 18).

If, on the other hand, the two outer lines, AB and EF, are both parallel to the middle line CD, then every line AK drawn from A into the angle $\angle$BAE will cut the line CD at some point K, regardless of how small the angle $\angle BAK$ is taken. Draw a line joining C to an arbitrary point L on the extension of AK. The line CL must cut EF at some point M, producing the triangle $\triangle MCE$. Since the extension of the line AL into the triangle $\triangle MCE$ can cut neither AC nor CM a second time, it must cut EF at some point H. Hence, AB and EF are mutually parallel.

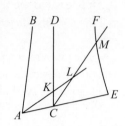

Suppose now that two parallels, AB and CD, lie in two planes whose line of intersection is EF. From an arbitrarily chosen point E of EF, drop a perpendicular EA to one of the parallels, say to AB. From the foot of this perpendicular, A, drop a new perpendicular, AC, to CD, the other parallel. Draw the line EC joining E and C, the endpoints of this perpendicular construction. The angle $\angle BAC$ must be acute (TP 22), so the foot G of a perpendicular CG dropped from C to AB will fall

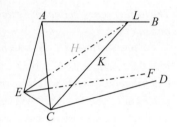

on that side of AC in which the lines AB and CD are parallel. The line EC, together with any line EH that enters angle $\angle AEF$ (regardless of how slightly EH deviates from EF), determines a plane. This plane must cut the plane of the parallels AB and CD along some line CK. This line cuts AB somewhere – namely, at the very point L common to all three planes, through which the line EH necessarily passes as well. Thus, EF is parallel to AB. We can establish the parallelism of EF and CD similarly.[3]

Therefore, a line EF is parallel to one of a pair of parallels, AB and CD, if and only if EF is the intersection of two planes, each containing one of the parallels, AB and CD. Thus, two lines are parallel to one another if they are parallel to a third line, even if the lines do not all lie in one plane. This last sentence could also be expressed thus: the lines in which three planes intersect must all be parallel to one another if the parallelism of two of the lines is established.

Proposition 26

Antipodal spherical triangles have equal areas.

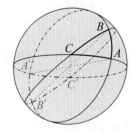

By antipodal triangles, I mean those triangles that are formed on opposite sides of a sphere when three planes through its center intersect it. It follows that antipodal triangles have their sides and angles in reverse order.

By antipodal triangles, I mean those triangles that are formed on opposite sides of a sphere when three planes through its center intersect it. It follows that antipodal triangles have their sides and angles in reverse order.

---

[3] For the sake of clarity, I have taken the liberty of changing the names of some of the points in this passage: the points that I have called H, K, and L are *all* called H in Lobachevski's original. $-SB$

The corresponding sides of antipodal triangles $\triangle ABC$ and $\triangle A'B'C'$ are equal: $AB = A'B'$, $BC = B'C'$, $CA = C'A'$. The corresponding angles are also equal: those at A, B, and C equal those at A', B', and C' respectively.

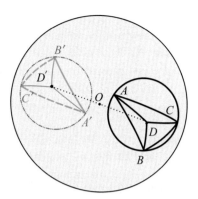

Consider the plane passing through the points A, B, and C. Drop a perpendicular to it from the center of the sphere, and extend this perpendicular in both directions; it will pierce the antipodal triangles in antipodal points, D and D'. The distances from D to the points A, B, and C, as measured along great circles of the sphere, must be equal, not only to one another (TP 12), but also to the distances $D'A'$, $D'B'$, and $D'C'$ on the antipodal triangle (TP 6). From this, it follows that the three isosceles triangles that surround D and comprise the spherical triangle $\triangle ABC$ are congruent to the corresponding isosceles triangles surrounding D' and comprising $\triangle A'B'C'$.

As a basis for determining when two figures on a surface are equal, I adopt the following postulate: two figures on a surface are equal in area when they can be formed by joining or detaching equal parts.

Proposition 27

A trihedral angle equals half the sum of its dihedral angles minus a right angle.

Let $\triangle ABC$ be a spherical triangle, each of whose sides is less than half a great circle. Let A, B, and C denote the measures of its angles. Extending side AB to a great circle divides the sphere into two equal hemispheres. In the one containing $\triangle ABC$, extend the triangle's other two sides through C, denoting their second intersections with the great circle by A' and B'. In this way, the hemisphere is split into four triangles: $\triangle ABC$, $\triangle ACB'$, $\triangle B'CA'$, and $\triangle A'CB$, whose sizes we shall denote by P, X, Y, and Z respectively.

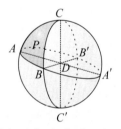

Clearly, $P + X = B$, and $P + Z = A$. Moreover, since the size Y of the spherical triangle $\triangle B'CA'$ equals that of its antipodal triangle $\triangle ABC'$ [TP 26], it follows that $P + Y = C$. Therefore, since $P + X + Y + Z = \pi$, we conclude that $P = \frac{1}{2}(A + B + C - \pi)$.

*

It is also possible to reach this conclusion by another method, based directly upon the postulate on equivalence of areas given above [in TP 26].

In the spherical triangle $\triangle ABC$, bisect the sides AB and BC, and draw the great circle through D and E, their midpoints. Drop perpendiculars AF, BH, and CG upon this circle from A, B, and C.

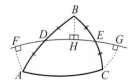

If H, the foot of the perpendicular dropped from B, falls between D and E, then the resulting right triangles $\triangle BDH$ and $\triangle AFD$ will be congruent, as will $\triangle BHE$ and $\triangle EGC$ (TP 6 & 15). From this, it follows that the area of triangle $\triangle ABC$ equals that of the quadrilateral $AFGC$.

If H coincides with E (see figure at left), only two equal right triangles will be produced, $\triangle AFD$ and $\triangle BDE$. Interchanging them establishes the equality of area of triangle $\triangle ABC$ and quadrilateral $AFGC$.

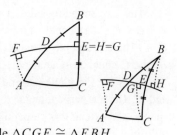

Finally, if H falls outside triangle $\triangle ABC$ (see figure at right), the perpendicular CG must enter the triangle. We may then pass from triangle $\triangle ABC$ to quadrilateral $AFGC$ by adding triangle $\triangle FAD \cong \triangle DBH$ and then taking away triangle $\triangle CGE \cong \triangle EBH$.

Since the diagonal arcs AG and CF of the spherical quadrilateral $AFGC$ are equal to one another (TP 15), the triangles $\triangle FAC$ and $\triangle ACG$ are congruent to one another (TP 15), whence the angles $\angle FAC$ and $\angle ACG$ are equal to one another. Hence, in all the preceding cases, the sum of the three angles in the spherical triangle equals that of the two equal, non-right angles in the quadrilateral.

Therefore, given any spherical triangle whose angle sum is S, there is a quadrilateral with two right angles of the same area, each of whose other two angles equals $S/2$.

Let $ABCD$ be such a quadrilateral, whose equal sides AB and DC are perpendicular to BC, and whose angles at A and D each equal $S/2$. Extend its sides AD and BC until they meet at E; extend AD beyond E to F, so that $EF = ED$, and then drop a perpendicular FG upon the extension of BC. Bisect the arc BG, and join its midpoint H to A and F with great circle arcs.

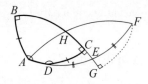

The congruence of the triangles $\triangle EFG$ and $\triangle DCE$ (TP 15) implies that $FG = DC = AB$. The right triangles $\triangle ABH$ and $\triangle HGF$ are also congruent, since their corresponding arms are equal. From this it follows that the arcs AH and AF belong to the same great circle. Thus, the arc AHF is half a great circle, as is the arc $ADEF$. Since $\angle HFE = \angle HAD = S/2 - \angle BAH = S/2 - \angle HFG = S/2 - \angle HFE - \angle EFG = S/2 - \angle HAD - \pi + S/2$, we conclude that $\angle HFE = \frac{1}{2}(S - \pi)$. Equivalently, we have shown that $\frac{1}{2}(S - \pi)$ is the size of the spherical lune $AHFDA$, which in turn equals the size of the quadrilateral ABCD; this last equality is easy to see, since we may pass from one to the other by first adding the triangles $\triangle EFG$ and $\triangle BAH$, and then removing triangles that are congruent to them: $\triangle DCE$ and $\triangle HFG$.

Therefore, $\frac{1}{2}(S - \pi)$ is the size both of the quadrilateral $ABCD$, and of the spherical triangle, whose angle sum is S.

Proposition 28

If three planes intersect one another along parallel lines, the sum of the three resulting dihedral angles is equal to two right angles.

Suppose that three planes intersect one another along three parallel lines, AA', BB', and CC' (TP 25). Let X, Y, and Z denote the dihedral angles they form at AA', BB', and CC', respectively. Take random points A, B, and C, one from each line, and construct the plane passing through them. Construct a second plane containing the line AC and some point D of BB'. Let the dihedral angle that this plane makes with the plane containing the parallel lines AA' and CC' be denoted by w.

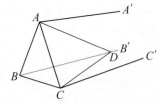

Draw a sphere centered at A; the points in which the lines AC, AD, and AA' intersect it determine a spherical triangle, whose size we shall denote by α, and whose sides we shall denote p, q, and r.

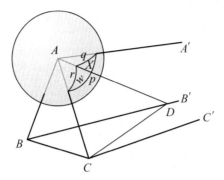

If q and r are those sides whose opposite angles have measures w and X respectively, then the angle opposite side p must have measure $\pi + 2\alpha gw - X$. (TP 27)

Similarly, the intersections of CA, CD, and CC' with a sphere centered at C determine a spherical triangle of size β, whose sides are denoted by p', q', and r', and whose angles are: w opposite q', Z opposite r', and thus, $\pi + 2\beta - w - Z$ opposite p'.

Finally, the intersections of DA, DB, and DC with a sphere centered at D determine a spherical triangle, whose sides, l, m, and n, lie opposite its angles, $w + Z - 2\beta$, $w + X - 2\alpha$, and Y, respectively. Its size, consequently, must be $\delta = \frac{1}{2}(X + Y + Z - \pi) - (\alpha + \beta - w)$.

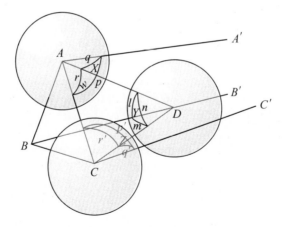

If w decreases toward zero, then α and β will vanish as well, so that $(\alpha + \beta gw)$ can be made less than any given number. Since sides l and m of triangle δ will also vanish (TP 21), we can, by taking w sufficiently small, place as many copies of δ as we wish, end to end, along the great circle containing m, without completely covering the hemisphere with triangles in the process. Hence, δ vanishes together with w. From this, we conclude that we must have $X + Y + Z = \pi$.

Proposition 29

In a rectilinear triangle, the three perpendicular bisectors of the sides meet either in a single point, or not at all.

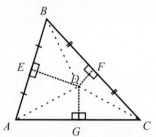

Suppose that two of triangle ABC's perpendicular bisectors, say, those erected at the midpoints E & F of AB and BC respectively, intersect at some point, D, which lies within the triangle. Draw the lines DA, DB, and DC, and observe that the congruence of the triangles ADE and BDE (TP 10) implies that $AD = BD$. For similar reasons, we have $BD = CD$, whence it follows that triangle ADC is isosceles. Consequently, the perpendicular dropped from D to AC must fall upon AC's midpoint, G.

This reasoning remains valid when D, the point of intersection of the two perpendiculars ED and FD, lies outside the triangle, or when it lies upon side AC.

Thus, if two of the three perpendiculars fail to intersect one another, then neither of them will intersect the third.

Proposition 30

In a rectilinear triangle, if two of the perpendicular bisectors of the sides are parallel, then all three of them will be parallel to one another.

In triangle $\triangle ABC$, erect perpendiculars DE, FG, and HK from D, F, and H, the midpoints of the sides. (See the figure.)

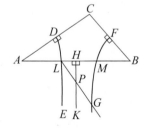

We first consider the case in which DE and FG are parallel, and the third perpendicular, HK, lies between them. Let L and M be the points in which the parallels DE and FG cut the line AB. Draw an arbitrary line entering angle $\angle$BLE through L. Regardless of how small an angle it makes with LE, this line must cut FG (TP 16); let G be the point of intersection. The perpendicular HK enters triangle $\triangle LGM$, but because it cannot intersect MG (TP 29), it must exit through LG at some point P. From this it follows that HK must be parallel to DE (TP 16 & 18) and FG (TP 18 & 25).

In the case just considered, if we let the sides $BC = 2a$, $AC = 2b$, $AB = 2c$, and denote the angles opposite them by A, B, C, we can easily show that

$$A = \Pi(b) - \Pi(c) \quad B = \Pi(a) - \Pi(c) \quad C = \Pi(a) + \Pi(b)$$

by drawing lines AA', BB', CC', from points A, B, C, parallel to HK— and therefore parallel to DE and FG as well (TP 23 & 25).

Next, consider the case in which HK and FG are parallel. Since DE cannot cut the other two perpendiculars (TP 29), it either is parallel to them, or intersects AA'. To assume this latter possibility is to assume that $C > \Pi(a) + \Pi(b)$.

If this is the case, we can decrease the magnitude of this angle to $\Pi(a) + \Pi(b)$ by rotating line AC to a new position CQ (see the figure). The angle at B is thereby increased. That is, in terms of the formula proved above,

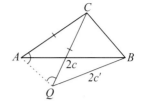

$$\Pi(a) - \Pi(c') > \Pi(a) - \Pi(c),$$

where $2c'$ is the length of BQ. From this it follows that $c' > c$ (TP 23).

On the other hand, since the angles at A and Q in triangle $\triangle ACQ$ are equal, the angle at Q in triangle $\triangle ABQ$ must be greater than the angle at A in the same triangle. Consequently, $AB > BQ$ (TP 9); that is, $c > c'$.

Proposition 31

We define a horocycle to be a plane curve with the property that the perpendicular bisectors of its chords are all parallel to one another.

In accordance with this definition, we may imagine generating a horocycle as follows: from a point A on a given line AB, draw various chords AC of length $2a$, where $\Pi(a) = \angle CAB$. The endpoints of such chords will lie on the horocycle, whose points we may thus determine one by one.

The perpendicular bisector DE of a chord AC will be parallel to the line AB, which we shall call the *axis* of the horocycle. Since the perpendicular bisector FG of any chord AH will be parallel to AB, the perpendicular bisector KL of any chord CH will be parallel to AB as well, regardless of the points C and H on the horocycle between which the chord is drawn (TP 30). For that reason, we shall not distinguish AB alone, but shall instead call *all* such perpendiculars *axes of the horocycle.*

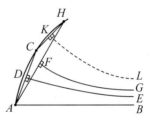

Proposition 32

A circle of increasing radius merges into a horocycle.

Let AB be a chord of the horocycle. From its endpoints, A and B, draw the two axes AC and BD; these will necessarily make equal angles, $BAC = ABD = \alpha$, with the chord AB (TP 31). From either axis, say AC, select an arbitrary point E to be the center of a circle. Draw an arc of this circle extending from A to F, the point at which it intersects BD. The circle's radius EF will make angle $AFE = \beta$ on one side of the chord of the circle, AF; on the other side, it will make angle $EFD = \gamma$ with the axis BD.

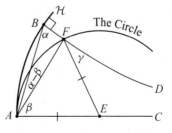

Now, angle γ will decrease if we move F toward B along axis BF while holding the center E fixed (TP 21). Moreover, γ will decrease to zero if we move the center E down axis AC while holding F fixed (TP 21, 22).

As γ vanishes, so does $\alpha - \beta$, the angle between AB and AF. Consequently, the distance from point B of the horocycle to point F of the circle vanishes as well. For this reason, one may also call the horocycle a circle of infinite radius.

Proposition 33

Let $AA' = BB' = x$ be segments of two lines that are parallel in the direction from A to A'. If these parallels are axes of two horocycles, whose arcs $AB = s$ and $A'B' = s'$ they delimit, then the equation $s' = se^{-x}$ holds, where e is some number independent of the arcs s, s', and the line segment x, the distance between the arcs s' and s.

Suppose that n and m are whole numbers such that $s : s' = n : m$. Draw a third axis CC' between AA' and BB'. Let $t = AC$ and $t' = A'C'$ be the lengths of the arcs that it cuts from AB and $A'B'$ respectively. Assuming that $t : s = p : q$ for some whole numbers p and q, we have

$$s = (n/m)s' \text{ and } t = (p/q)s.$$

If we divide s into nq equal parts by axes, any one such part will fit exactly mq times into s' and exactly np times into t. At the same time, the axes dividing s into nq equal parts divide s' into nq equal parts as well. From this it follows that

$$t'/t = s'/s.$$

Consequently, as soon as the distance x between the horocycles is given, the ratio of t to t' is determined; this ratio remains the same, no matter where we draw CC' between AA' and BB'.

From this, it follows that if we write $s = es'$ when $x = 1$, then $s' = se^{-x}$ for every value of x.

We may choose the unit of length with which we measure x as we see fit. In fact, because e is an undetermined number subject only to the condition $e > 1$, we may, for the sake of computational ease, choose the unit of length so that the number e will be the base of the natural logarithm.

In addition, since $s' = 0$ when $x = \infty$, we observe that, in the direction of parallelism, the distance between two parallels not only decreases (TP 24), but ultimately vanishes. Thus, parallel lines have the character of asymptotes.

Proposition 34

We define a horosphere to be the surface generated by revolving a horocycle about one of its axes, which, together with all the remaining axes of the horocycle, will be an axis of the horosphere.

Any chord joining two points of the horosphere will be equally inclined to the axes that pass through its endpoints, regardless of which two points are taken.

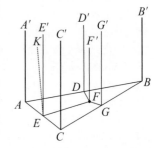

Let A, B, and C be three points on the horosphere, where AA' is the axis of rotation and BB' and CC' are any other axes. The chords AB and AC will be equally inclined toward the axes passing through their endpoints; that is, $\angle A'AB = \angle B'BA$ and $\angle A'AC = \angle C'CA$ (TP 31). The axes BB' and CC' drawn through the endpoints of the third chord BC are, like those of the other chords, parallel and coplanar with one another (TP 25).

The perpendicular DD' erected from the midpoint D of chord AB in the plane of the two parallels AA', BB' must be parallel to the three axes AA', BB', CC' (TP 31, 25). Similarly, the perpendicular bisector EE' of chord AC in the plane of parallels AA', CC' will be parallel to the three axes AA', BB', CC', as well as the perpendicular bisector DD'.

Denote the angle between the plane of the parallels AA', BB' and the plane in which triangle $\triangle ABC$ lies by $\Pi(a)$, where a may be positive, negative, or zero. If a is positive, draw $DF = a$ in the plane of triangle $\triangle ABC$, into the triangle, perpendicular to chord AB at its midpoint D; if a is negative, draw $DF = a$ outside the triangle on the other side of chord AB; if $a = 0$, let point F coincide with D.

All cases give rise to two congruent right triangles, $\triangle AFD$ and $\triangle DFB$, whence $FA = FB$. From F, erect FF' perpendicular to the plane of triangle $\triangle ABC$.

Because $\angle D'DF = \Pi(a)$ and $DF = a$, FF' must be parallel to DD'; the plane containing these lines is perpendicular to the plane of triangle $\triangle ABC$.

Moreover, FF' is parallel to EE'; the plane containing them is also perpendicular to the plane of triangle $\triangle ABC$.

Next, draw EK perpendicular to EF in the plane containing the parallels EE' and FF'. It will be perpendicular to the plane of triangle $\triangle ABC$ (TP 13), and hence to the line AE lying in this plane. Consequently, AE, being perpendicular to EK and EE', must be perpendicular to FE as well (TP 11). The triangles $\triangle AEF$ and $\triangle CEF$ are congruent, since they each have a right angle, and their corresponding sides about their right angles are equal. Therefore, $FA = FC = FB$.

In isosceles triangle $\triangle BFC$, a perpendicular dropped from vertex F to the base BC will fall upon its midpoint G.

The plane containing FG and FF' will be perpendicular to the plane of triangle $\triangle ABC$, and will cut the plane containing the parallels BB', CC' along a line that is parallel to them, GG'. (TP 25).

Since CG is perpendicular to FG, and thus to GG' as well [TP 13], it follows that $\angle C'CG = \angle B'BG$ (TP 23).

From this, it follows that any axis of the horosphere may be considered its axis of rotation.

We shall refer to any plane containing an axis of a horosphere as a *principal plane*. The intersection of the principal plane with the horosphere is a horocycle; for any other cutting plane, the intersection is a circle.

Any three principal planes that mutually cut one another will meet at angles whose sum is π (TP 28). We shall consider these the angles of a *horospherical triangle*, whose sides are the arcs of the horocycles in which the three principal planes intersect the horosphere. Accordingly, *the relations that hold among the sides and angles of horospherical triangles are the very same that hold for rectilinear triangles in the ordinary geometry.*

Proposition 35

In what follows, we shall use an accented letter, *e.g.* x', to denote the length of a line segment when its relation to the segment which is denoted by the same, but unaccented, letter is described by the equation $\Pi(x) + \Pi(x') = \pi/2$.

Let $\triangle ABC$ be a rectilinear right triangle, where the hypotenuse is $AB = c$, the other sides are $AC = b$, $BC = a$, and the angles opposite them are $\angle BAC = \Pi(\alpha)$, $\angle ABC = \Pi(\beta)$. At point A, erect the line AA', perpendicular to the plane of triangle $\triangle ABC$; from B and C, draw BB' and CC' parallel to AA'.

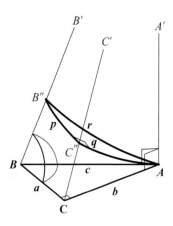

The planes in which these parallels lie meet one another at the following dihedral angles: $\Pi(\alpha)$ at AA', a right angle at CC' (TP 11 & 13), and therefore, $\Pi(\alpha')$ at BB' (TP 28).

The points at which the lines BB', BA, BC intersect a sphere centered at B determine a spherical triangle $\triangle mnk$, whose sides are $mn = \Pi(c)$, $kn = \Pi(\beta)$, $mk = \Pi(a)$, and whose opposite angles are, respectively, $\Pi(b)$, $\Pi(\alpha')$, $\pi/2$.

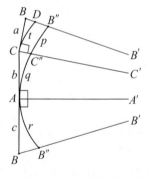

Thus, the existence of a rectilinear triangle with sides a, b, c and opposite angles $\Pi(\alpha)$, $\Pi(\beta)$, $\pi/2$ implies the existence of a spherical triangle with sides $\Pi(c)$, $\Pi(\beta)$, $\Pi(a)$ and opposite angles $\Pi(b)$, $\Pi(\alpha')$, $\pi/2$.

Conversely, the existence of such a spherical triangle implies the existence of such a rectilinear triangle.

Indeed, the existence of such a spherical triangle also implies the existence of a second rectilinear triangle, with sides a, α', β and opposite angles $\Pi(b')$, $\Pi(c)$, $\pi/2$. Hence, we may pass from a, b, c, α, β to b, a, c, β, α, and to a, α', β, b', c, as well.

If the horosphere through A with axis AA' cuts BB' and CC' at B'' and C'', its intersections with the planes formed by the parallels produce a horospherical triangle with sides $B''C'' = p$, $C''A = q$, $B''A = r$ and opposite angles $\Pi(\alpha')$, $\Pi(\alpha)$, $\pi/2$.

Consequently (TP 34),

$$p = r \sin \Pi(\alpha) \text{ and } q = r \cos \Pi(\alpha).$$

Along BB', break the connection of the three principal planes, turning them out from one another so that they lie in a single plane. In this plane, the arcs p, q, r unite into an arc of a single horocycle, which passes through A and has axis AA'.

Thus, the following lie on one side of AA': arcs p and q; side b of the rectilinear triangle, which is perpendicular to AA' at A; axis CC', which emanates from the endpoint of b, then passes through C'', the join of p and q, and is parallel to AA'; and the axis BB', which emanates from the endpoint of a, then passes through B'', the endpoint of arc p, and is parallel to AA'. On the other side of AA' lie the following: side c, which is perpendicular to AA' at point A, and axis BB', which emanates from the endpoint of c, then passes through B'', the endpoint of arc r, and is parallel to AA'.

Moreover, we see (by TP 33) that

$$t = pe^{f(b)} = r \sin \Pi(\alpha)e^{f(b)}.$$

If we were to erect the perpendicular to triangle $\triangle ABC$'s plane at B, instead of A, then the lines c and r would remain the same, while the arcs q and t would change to t and q, the straight lines a and b would change to b and a, and the angle $\Pi(\alpha)$ would change to $\Pi(\beta)$. From this it follows that

$$q = r \sin \Pi(\beta)e^{f(a)}.$$

Thus, by substituting the value that we previously obtained for q, we find that

$$\cos \Pi(\alpha) = \sin \Pi(\beta) e^{f(a)}.$$

If we change α and β into b' and c, then

$$\sin \Pi(b) = \sin \Pi(c) e^{f(a)}.$$

Multiplying by $e^{f(b)}$ yields

$$\sin \Pi(b) e^{f(b)} = \sin \Pi(c) e^{f(c)}.$$

Consequently, it follows that

$$\sin \Pi(a) e^{f(a)} = \sin \Pi(b) e^{f(b)}.$$

Because the lengths a and b are independent of one another and, moreover, $f(b) = 0$ and $\Pi(b) = \pi/2$ when $b = 0$, it follows that for every a,

$$e^{-f(a)} = \sin \Pi(a).$$

Therefore,

$$\sin \Pi(c) = \sin \Pi(a) \sin \Pi(b)$$
$$\sin \Pi(\beta) = \cos \Pi(\alpha) \sin \Pi(a).$$

Moreover, by transforming the letters, these equations become

$$\sin \Pi(\alpha) = \cos \Pi(\beta) \sin \Pi(b)$$
$$\cos \Pi(b) = \cos \Pi(c) \cos \Pi(\alpha)$$
$$\cos \Pi(a) = \cos \Pi(c) \cos \Pi(\beta).$$

In the spherical right triangle, if the sides $\Pi(c)$, $\Pi(\beta)$, $\Pi(a)$ and opposite angles $\Pi(b)$, $\Pi(\alpha')$ are renamed a, b, c, A, B, respectively, then the preceding equations will assume forms that are known as established theorems of the ordinary spherical trigonometry of right triangles. Namely,

$$\sin(a) = \sin(c) \sin(A)$$
$$\sin(b) = \sin(c) \sin(B)$$
$$\cos(B) = \cos(b) \sin(A)$$
$$\cos(A) = \cos(a) \sin(B)$$
$$\cos(c) = \cos(a) \cos(b).$$

From these equations, we may derive those for all spherical triangles in general. Consequently, the formulae of spherical trigonometry do not depend upon whether or not the sum of the three angles in a rectilinear triangle is equal to two right angles.

Proposition 36

We now return to the rectilinear right triangle $\triangle ABC$ with sides a, b, c, and opposite angles $\Pi(\alpha), \Pi(\beta), \pi/2$. Extend this triangle's hypotenuse beyond B to a point D at which $BD = \beta$, and erect a perpendicular DD' from BD. By construction, DD' is parallel to BB', the extension of side a beyond B. Finally, draw AA' parallel to DD'; it will be parallel to CB' as well (TP 25).

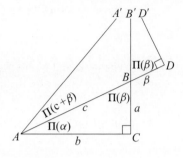

From this, we have $\angle A'AC = \Pi(b)$ and $\angle A'AD = \Pi(c + \beta)$, from which it follows that

$$\Pi(b) = \Pi(\alpha) + \Pi(c + \beta).$$

Now let E be the point on ray BA for which $BE = \beta$. Erect the perpendicular EE' to AB, and draw AA'' parallel to it. Line BC', the extension of side a beyond C, will be a third parallel.

If $\beta < c$, as in the figure, we see that $\angle CAA'' = \Pi(b)$ and $\angle EAA'' = \Pi(c - \beta)$, from which it follows that

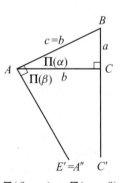

$$\Pi(c - \beta) = \Pi(\alpha) + \Pi(b).$$

In fact, this last equation remains valid even when $\beta = c$, or $\beta > c$.

If $\beta = c$ (see figure at left), the perpendicular AA' erected upon AB is parallel to BC, and hence to CC', from which it follows that $\Pi(\alpha) + \Pi(b) = \pi/2$. Moreover, $\Pi(c - \beta) = \pi/2$ (TP 23).

If $\beta > c$ (see figure at right), E falls beyond point A. In this case, we have $\angle EAA'' = \Pi(c - \beta)$, from which it follows that

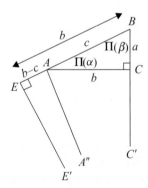

$$\Pi(\alpha) + \Pi(b) = \pi - \Pi(\beta - c) = \Pi(c - \beta) \quad \text{(TP 23)}.$$

Combining the two equations yields

$$2\Pi(b) = \Pi(c - \beta) + \Pi(c + \beta)$$
$$2\Pi(\alpha) = \Pi(c - \beta) - \Pi(c + \beta),$$

from which follows

$$\frac{\cos \Pi(b)}{\cos \Pi(\alpha)} = \frac{\cos[\frac{1}{2}\Pi(c - \beta) + \frac{1}{2}\Pi(c + \beta)]}{\cos[\frac{1}{2}\Pi(c - \beta) - \frac{1}{2}\Pi(c + \beta)]}.$$

Using the substitution

$$\frac{\cos \Pi(b)}{\cos \Pi(\alpha)} = \cos \Pi(c) \text{ (from TP 35)}$$

yields

$$\tan^2 \left(\frac{\Pi(c)}{2} \right) = \tan \left(\frac{\Pi(c - \beta)}{2} \right) \tan \left(\frac{\Pi(c + \beta)}{2} \right).$$

Because the angle $\Pi(\beta)$ at B may have any value between 0 and $\pi/2$, β itself can be any number between 0 and ∞. By considering the cases in which $\beta = c, 2c, 3c$, etc., we may deduce that for all positive values of r,[4]

$$\tan^r \left(\frac{\Pi(c)}{2} \right) = \tan \left(\frac{\Pi(rc)}{2} \right).$$

If we view r as the ratio of two values x and c, and assume that $\cot(\Pi(c)/2) = e^c$, we find that for all values of x, whether positive or negative,

$$\tan \left(\frac{\Pi(x)}{2} \right) = e^{-x},$$

where e is an indeterminate constant, which is larger than 1, since $\Pi(x) = 0$ when $x = \infty$.

Since the unit with which we measure lengths may be chosen at will, we may choose it so that e is the base of the natural logarithm.

Proposition 37

Of the five equations above (TP 35), the following two

$$\sin \Pi(c) = \sin \Pi(a) \sin \Pi(b)$$
$$\sin \Pi(\alpha) = \cos \Pi(\beta) \sin \Pi(b)$$

suffice to generate the other three: we can obtain one of the others by applying the second equation to side a rather than side b; we then deduce another by combining the equations already established. There will be no ambiguities of algebraic sign, since all angles here are acute.

Similarly, we obtain the two equations:

(1) $\tan \Pi(c) = \sin \Pi(\alpha) \tan \Pi(a)$

(2) $\cos \Pi(a) = \cos \Pi(c) \cos \Pi(\beta)$.

We shall now consider an arbitrary rectilinear triangle with sides a, b, c and opposite angles A, B, C.

If A and B are acute angles, then the perpendicular p dropped from C will fall within the triangle and cut side c into two parts: x, on the side of A, and $c - x$, on the side of B. This produces two right triangles. Applying equation (1) to each yields

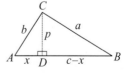

---

[4] Where I have r, Lobachevski uses the symbol n. Whatever one calls it, it stands for any positive *real* number. I have switched to r so as to conform with the convention of reserving n for natural numbers.

$$\tan \Pi(a) = \sin(B) \tan \Pi(p).$$

$$\tan \Pi(b) = \sin(A) \tan \Pi(p).$$

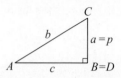

These equations hold even if one of the angles, say B, is right or obtuse. Thus, for any rectilinear triangle whatsoever, we have

(3) $\sin(A) \tan \Pi(a) = \sin(B) \tan \Pi(b).$

Applying equation (2) to a triangle with acute angles at A and B yields

$$\cos \Pi(x) = \cos(A) \cos \Pi(b)$$

$$\cos \Pi(c - x) = \cos(B) \cos \Pi(a).$$

These equations hold even when one of the angles A or B is right or obtuse.

For instance, when $B = \pi/2$, we have $x = c$; in this case, the first equation reduces to equation (2) and the second is trivially true. When $B > \pi/2$, applying equation (2) still yields the first equation; in place of the second, it yields $\cos \Pi(x - c) = \cos(\pi - B) \cos \Pi(a)$, which, however, is identical to the second, since $\cos \Pi(x - c) = -\cos \Pi(c - x)$ (TP 23), and $\cos(\pi - B) = -\cos(B)$. Finally, if A is right or obtuse, we must use $c - x$ and x, instead of x and $c - x$, so that the two equations will hold in this case also.

To eliminate x from the two equations above, we observe that

$$
\cos \Pi(c - x) = \frac{1 - \left[\tan\left(\frac{\Pi(c-x)}{2} \right) \right]^2}{1 + \left[\tan\left(\frac{\Pi(c-x)}{2} \right) \right]^2}
$$

$$
= \frac{1 - e^{2x - 2c}}{1 + e^{2x - 2c}}
$$

$$
= \frac{1 - \left[\tan\left(\frac{\Pi(c)}{2} \right) \right]^2 \left[\cot\left(\frac{\Pi(x)}{2} \right) \right]^2}{1 + \left[\tan\left(\frac{\Pi(c)}{2} \right) \right]^2 \left[\cot\left(\frac{\Pi(x)}{2} \right) \right]^2}
$$

$$
= \frac{\cos \Pi(c) - \cos \Pi(x)}{1 - \cos \Pi(c) \cos(x)}
$$

If we substitute the expressions for $\cos \Pi(x)$ and $\cos \Pi(c - x)$ into this, it becomes from which it follows that

$$
\cos \Pi(a) \cos(B) = \frac{\cos \Pi(c) - \cos \Pi(A) \cos \Pi(b)}{1 - \cos \Pi(A) \cos \Pi(b) \cos \Pi(c)}, \quad {}_5
$$

---

[5] In Lobachevski's original, the positions of this equation and the preceding one are reversed: presumably this was a printer's error. The fact that Halsted perpetuated it in his 1891 translation leads me to suspect that Halsted, realizing that one could reach the conclusions of TP 37 by simpler arguments than Lobachevski's own, did not bother to look very closely at the details as they stand.

and finally,

(4) $[\sin \Pi(c)]^2 = [1 - \cos(B)\cos \Pi(c)\cos \Pi(a)]\,[1 - \cos(A)\cos \Pi(b)\cos \pi(c)]\,.$

Similarly, we also have

$$[\sin \Pi(a)]^2 = [1 - \cos(C)\cos \Pi(a)\cos \Pi(b)]\,[1 - \cos(B)\cos \Pi(c)\cos \Pi(a)]$$
$$[\sin \Pi(b)]^2 = [1 - \cos(A)\cos \Pi(b)\cos \Pi(c)]\,[1 - \cos(C)\cos \Pi(a)\cos \Pi(b)]\,.$$

From these three equations, we find that

$$\frac{[\sin \Pi(b)]^2\,[\sin \Pi(c)]^2}{[\sin \Pi(a)]^2} = [1 - \cos(A)\cos \Pi(b)\cos \Pi(c)]^2\,.$$

From this it follows, without ambiguity of sign, that

(5) $\cos(A)\cos \Pi(b)\cos \Pi(c) + \dfrac{\sin \Pi(b)\sin \Pi(c)}{\sin \Pi(a)} = 1.$

The following expression for $\sin \Pi(c)$ follows from an alternate form of (3):

$$\sin \Pi(c) = \frac{\sin(A)}{\sin(C)}\tan \Pi(a)\cos \Pi(c).$$

If we substitute this expression into equation (5), we obtain

$$\cos \Pi(c) = \frac{\cos \Pi(a)\sin(C)}{\sin(A)\sin \Pi(b) + \cos(A)\sin(C)\cos \Pi(a)\cos \Pi(b)}.$$

If we substitute this expression for $\cos \Pi(c)$ into equation (4), we obtain

(6) $\cot(A)\sin(C)\sin \Pi(b) + \cos(C) = \dfrac{\cos \Pi(b)}{\cos \Pi(a)}.$

By eliminating $\sin \Pi(b)$ with the help of equation (3), we find that

$$\frac{\cos \Pi(a)}{\cos \Pi(b)}\cos(C) = 1 - \frac{\cos(A)}{\sin(B)}\sin(C)\sin \Pi(a).$$

On the other hand, permuting the letters in equation (6) yields

$$\frac{\cos \Pi(a)}{\cos \Pi(b)} = \cot(B)\sin(C)\sin \Pi(a) + \cos(C).$$

By combining the last two equations, we obtain

(7) $\cos(A) + \cos(B)\cos(C) = \dfrac{\sin(B)\sin(C)}{\sin \Pi(a)}.$

Thus, the four equations that describe how the sides a,b,c and angles A, B, C are interrelated in rectilinear triangles are [equations (3), (5), (6), (7)]:

$$(8) \quad \begin{cases} \sin(A)\tan \Pi(a) = \sin(B)\tan \Pi(b) \\[2mm] \cos(A)\cos \Pi(b)\cos \Pi(c) + \dfrac{\sin \Pi(b)\sin \Pi(c)}{\sin \Pi(a)} = 1 \\[2mm] \cot(A)\sin(C)\sin \Pi(b) + \cos(C) = \dfrac{\cos \Pi(b)}{\cos \Pi(a)} \\[2mm] \cos(A) + \cos(B)\cos(C) = \dfrac{\sin(B)\sin(C)}{\sin \Pi(a)} \end{cases}$$

When the sides a, b, c of the triangle are very small, we may content ourselves with the following approximations (TP 36):

$$\cot \Pi(a) = a,$$
$$\sin \Pi(a) = 1 - \frac{1}{2}a^2,$$
$$\cos \Pi(a) = a,$$

where the same approximations hold for sides b and c also.

In the scholarly journal of the University of Kazan, I have published several investigations into the measurements of curves, plane figures, surfaces, and solids, as well as the application of imaginary geometry to analysis.

In and of themselves, the equations (8) already constitute sufficient grounds for believing that the imaginary geometry might be possible. As a result, we have no means other than astronomical observations with which to judge the accuracy that follows from calculations in the ordinary geometry. Its accuracy is very far-reaching, as I have demonstrated in one of my investigations; for example, in all angles who sides we are capable of measuring, the sum of the three angles does not differ from π by so much as a hundredth of a second.

Finally, it is worth observing that the four equations (8) of plane geometry become valid formulae of spherical geometry if we substitute $a\sqrt{-1}, b\sqrt{-1}, c\sqrt{-1}$ for the sides a, b, c ; these substitutions will change

$$\begin{array}{lcl} \sin \Pi(a) & \text{to} & \dfrac{1}{\cos a}, \\[3mm] \cos \Pi(a) & \text{to} & \sqrt{-1}\tan a. \\[3mm] tan\Pi(a) & \text{to} & \dfrac{1}{\sqrt{-1}\sin a}, \end{array}$$

and similarly for sides b and c. Hence, these substitutions change equations (8) into the following:

$$\sin A \sin b = \sin B \sin a$$
$$\cos a = \cos b \cos c + \sin b \sin c \cos A$$
$$\cot A \sin C + \cos C \cos b = \sin b \cot a$$
$$\cos A = \cos a \sin B \sin C - \cos B \cos C.$$

Index

About the Author

Seth Braver (B.A. San Francisco State University, M.A. University of California – Santa Cruz, Ph.D. University of Montana) was born in Atlanta, Georgia. He taught at the University of Montana (where he won several teaching awards) and St. John's College in Santa Fe (where he led classes in literature, philosophy, and ancient Greek, as well as mathematics) before joining the mathematics department of South Puget Sound Community College in Olympia, Washington, where he currently teaches.